Protected Adriatic

Author: Fabio VALLAROLA
Review and layout: Graziano ARETUSI
English translation: Mario CIPOLLONE

Publication as an e-book on Lulu platform
Printable text on demand on www.lulu.com and www.amazon.com

Scientific work produced within the PhD "Analysis of Development Policies and Promotion of the Territory" 2008-2011.
Prof. Bernardo CARDINALE - University of Teramo.

Free availability for the present publication.

Publication produced as part of the "Small Project" called "AdriaPAN Secretariat", financed by MEDPAN with funds FFEM, MAVA Foundation and the Foundation Prince Albert II of Monaco - 2nd call 2012-2013.

Dissemination and distribution within the project PANforAMaR – Protected Areas Network for Adriatic and Ionian Macro Region - funded by AII – Ionian and Adriatic Initiative – under the support cooperation programme – 1st call 2012-2014.

ISBN 978-1-291-56708-3
All rights reserved to the Author, 2013 ©

INDEX

PART I: MANAGEMENT TOOLS	**3**
1. The International framework and the EU Law	**4**
2. Tourism: useful resource but difficult to be managed	**6**
3. The situation in the Mediterranean and Adriatic	**13**
3.1. Fishing	18
3.2. Topic of supranational importance	21
3.3. The selection of marine and coastal protected areas	23
4. Planning of Marine Protected Areas	**25**
4.1. Planning and management of Marine Protected Areas	27
4.2. Integrated Coastal Zone Management	32
5. Programmes, budgets and participation	**36**
5.1. Participation and involvement	38
PART II: THE SITUATION IN THE ADRIATIC REGION	**43**
1. The Adriatic Region and the cooperation	**44**
1.1. The Adriatic ecosystem	47
1.2. The economic situation in the Adriatic countries	51
1.3. Cross-border cooperation	58
1.4. The Adriatic-Ionian Initiative	64
2. The Adriatic protected areas	**67**
2.1. Bibliographic and cartographic research	70
2.2. Direct surveys based on interviews	75
3. Working in a network: MEDPAN and AdriaPAN	**79**
3.1. Networks among protected areas	81
3.2. La MedPAN Network	84
3.3. La AdriaPAN Network	86
3.4. The added value of Networks	94
4. A survey on the Adriatic	**96**
Conclusions	**103**
Annex – "Cerrano" Charter	**105**
Bibliografy	**111**

PART I: MANAGEMENT TOOLS

1. The International framework and the EU Law

Every nation has its own view on how to consider protected areas in an overall planning. The various trends can be gathered in three strands. Strand One is that, for example followed by Denmark, which tends not to relate any special tool only to particular areas. Protected areas, i.e., are included within the more general land management tools such as landscape, natural or forestry planning. Strand Two stands for integration through special connection tools as it occurs in Germany and to some extent in England. Strand Three, instead, mainly followed also in Italy, is to equip the protected areas of specific authorities with their own management and planning tools, even partially autonomous and independent from those management guidelines and urban, territorial or land-use planning of which other local governments are in charge.

In the cultural debate on the specific issue of the regional planning, the need to establish protected areas, with their own borders and superior laws to ordinary ones, is considered a sort of inability of the modern urban science to control the soil-consuming economic forces and social dynamics. This debate occurs both at local and international levels.

Since the 70^s the European Community has started a European environmental management plan involving all countries in the application of common directives.

In the field of nature conservation it has been trying to develop over time a system in order to overcome the paradox of environmental protection implemented in the exact level and only in specific protected areas, rather than acting on the wide area (Marchisio, 1999).

The most successful initiative in the member states is the one following a series of Directives addressed at first to the protection of migratory birds, then also to habitats and other species of particular importance in the European territory.[1]

This is the programme called Natura 2000[2] aiming to create a European network of protected areas, identified generically as "Sites of Community Importance"[3], sometimes just to temper the situation of a conservation of nature made only by "islands of happiness". The identification of sites was carried out in Italy by every single region, although in a centrally coordinated process, providing there a major stimulus to the policy of nature conservation.

Thus, the way has been paved for a very positive relationship that is going on after the phase of recognition of sites in subsequent actions for the protection, management and activation of plans and projects for sustainable development.

In Italy the proposed sites were initially 2,413. Over time the number has increased to 2,565 and continues to grow for the need felt by the European Commission to identify a greater number of marine sites for all

---

[1] This refers in particular to two EU Directives: the Directive 79/409/CEE *"Birds"* of April 2, 1979 and the EU Directive 92/43/CEE *"Habitat"* of May 21, 1992.

[2] *Natura 2000* is the name that the Council of Ministers of the European Union gave to a coordinated and coherent system of areas for the conservation of biodiversity in the territory of the EU, and in particular for the conservation of a number of habitats, plant and animal species mentioned in the "Birds" and "Habitats" Directives. Source:
http://ec.europa.eu/environment/nature/natura2000/index_en.htm (01.02.2010).

[3] In fact the Natura 2000 programme refers to both the Sites of Community Interest (SCI) identified in compliance with the so-called Habitats Directive, and also the Special Protection Areas (SPAs) instead identified on the basis of criteria specified by the Directive "Birds". At international level, the abbreviations used are respectively: SCIs (Sites of Community Importance) and SPAs (Special Protection Areas).

European countries and Italy[4] in particular. At first the sites mainly coincided with the areas already under regional or national protective measures, for a variety of reasons related to the fact that certainly the most delicate areas were already under an order of protection at the time the project BioItaly started, but also because analysis, to be conducted necessarily in a undisturbed way to produce credible data, could be derived and made mainly in geographical areas where a form of protection was already guaranteed.

Today, the network of protected areas part of Natura 2000, SCI and SPAs, appears to be the discriminating requisite to apply for the EU funding for environmental projects. To be out of it means to have precluded many potential access to funding programmes such as the following two: Life Nature and Life plus, which annually continue to provide good possibilities to public administrations, businesses, associations and individuals on projects related to the conservation of species and habitats identified as priorities.

2. Tourism: useful resource but difficult to be managed

Ranging from the Western countries to the less developed continents, it is in place a complex process of community disintegration and loss of identity caused by the compelling globalization, aspects of the process of emancipation and democratization of societies, and resulting processes of rampant deregulation and uniformities. This worrying situation amounts, on the one hand, to the risk of a total loss of shared values that tie down people to communities and communities to their territory and, on the other

---

[4] Source: EU – European Commission, *Barometer Natura 2000,* Newsletter of DG Environment of the European Commission, no. 33, January 2013.

hand, to insecurity anxiety and discomfort that are likely to result in local bias and closing attitudes.

The modernity of Western culture favored the triumph of a very particular form of rationality in which the excessive organization of natural and human resources leads to an ever more complex and efficient society, in which the relationships tend to become more formal and impersonal. An ambivalent modernity, as identified by many thinkers at the turn of the nineteenth and twentieth centuries. On the one hand, it sees a universalization, that is a reason shared by all human beings of the world where the new surpasses the old and in which economic efficiency should be extended to all countries, and, on the other hand, a concept of dominant individualization that leads the individual to autonomy, independence and freedom of choice but also to lack of creativity, uniformity of behaviours, to a dehumanization of relationships and a complete isolation, "socially out of a community, however physically immersed in it" (De Marchi et al, 2001).

Still about sixty years ago the whole Adriatic region appeared as a largely rural society, with an industrial development only in the north-west part of Italy, despite to the rest of the advanced countries of the European continent. In all Adriatic nations two processes, usually closely connected but not necessarily ordered according to the same chronological scans, have intertwined: industrial development and social modernization. Along with demographic regimes based on low mortality, disappearance of great deficiency syndromes and infectious diseases (malaria, rickets, scurvy, etc.)., urbanization, rising rates of schooling, the modernization has been characterized by a progressive disappearance of rural society, with a migration of population not only from countryside to city, but also from agricultural labour to industry and services. In the 50^s rural society resisted in Italy as a social and economic context in which half of Italians lived and worked (De Bernardi and Ganapini, 1996).

Since 1992 the recommendations produced as part of the Earth Summit in Rio de Janeiro, through *Agenda 21*, up to the 5th program of Action for Sustainable Development, all official documents of the western countries have referred to the need to start a form of development, sustainable both for the ecosystem and the various social realities in it.[5]

The word *sustainable* has appeared for some time everywhere. If in other fields it may be a mere pretext to legitimize a more or less morally correct initiative, in the tourism industry it is a real challenge.

In fact, too often in this branch very different realities – different both in an environmental sense, intended as a possible conflict between the work of man and the conservation of nature, and in the social sense, when the explosive welfare of certain populations of the world alters the delicate balance of less developed continents - come into direct contact.

The greatest socio-political debate of this century has arisen around these topics. Two views stand opposite: on the one hand there are those who believe that the ongoing process of globalization is a way to bring democracy, freedom and prosperity to those countries where these features are often missing. On the other hand, there are those who believe that just this process prevents those countries to have a normal evolution towards better living conditions.

In this controversy, tourism can be one of the warmest "frontlines". Sometimes trying to make it "sustainable" is a hard task. In the just outlined debate, what appears to some people as a form of aid to the least developed countries, for others it can be understood as a form of colonization.

---

[5] *Brundtland Report of the World Commission on Environment and Development*, says: «the sustainable development is a development that meets the needs of the present without compromising the ability of future generations to meet their own needs».

It is estimated that tourism moves people in numbers of around 3-6 billion arrivals and overnight stays in the different countries of the world. Since 1980 it has started an international phenomenon: the emergence of new tourist destinations in countries where the number of visitors was not a traditional form of economic activity. This has led to a globalization of tourism which today affects every corner of the planet (Mercury, 2009).

We are witnessing in recent years the growth of a kind of tourism based on the enjoyment of nature. The protagonists of this tourism are people looking at nature not only for moments of regeneration but also for their own cultural growth. Protected areas can and should not only respond to this type of tourism, but also increase the respect of nature through conservation actions, education, compatible enjoyment for the very preservation of their natural heritage. For these reasons, the protected areas should operate to guide and qualify the tourist flows, so that the tourist organization becomes more qualified and typified.

To reach better and better functional efficiency conditions, through services, equipment and targeted interventions, is one of the objectives to be pursued by directing this action to a sustainable socio-economic development, made possible by intrinsic resources and characteristics of the territory.

This is the step where planning gives way to detailed design, a design which should be accompanied by a handbook that clarifies the concept of *genius loci*, often confused with a hypocrite stylistic traditionalism with no real adherence to any local identity.

On the contrary, interventions aimed at a proper use of the territory as a resource lead to the growth of the tourist industry on the one hand, on the other to the development of employment through the creation of new professionals. In addition to traditional jobs (for example those related to the implementation of nature trails or the creation of visitor centers through the renovation of existing historic buildings), the development of the culture of the place results in the creation of new activities oriented to an educational approach to the landscape.

In ancient times the community possessed formulas of sustainable balance, passed down by ancestral traditions such as the cult of the dead and the earth, focusing more attention and care for the "ground under our feet". It was thus not only a reason dictated by a cultural tradition connected to a primarily based on agriculture economy, nor superstition, but rather as the result of family education related to the worship of their ancestors to the earth, an element that produces what is needed to the sustenance of life, and that welcomes everybody in the end.

The temporal dimension is changed. However, the common consciousness of a resident community is always as stronger and more rooted to the spot as the reason for its existence is evident in it. It is precisely this reason that makes the *genius loci* more actual than ever as a radical element for sustainable development of tourism. It is a development that is always based on a perception of a territorial identity, in which people identifies before any external phenomenon or action to be taken in the future (Cestari, 2007).

The phenomenon of tourism is changing continuously and rapidly, with the steady growth and the reduction of important branches. Today a special interest seems to be focused on "active" holidays, involving a certain level of awareness and responsibility on the part of the tourist. This type of holiday coincides with a deconstruction of time, so that many holidays are programmed off-season and on several occasions, with a shortening of their time. In addition, the type of environmental and human impact of tourism, which has been progressively considered as an instrument of regional development and conservation of natural and cultural heritage, has become increasingly an object of consideration by the experts of the industry.

Tourism is a highly dynamic business and tourists are individuals with dynamic tastes. What attracts a traveler not necessarily attracts another. Someone likes to see some famous attractions, buy souvenirs, eat well and sleep in a comfortable hotel. Others are more interested in unknown and out-of-the-way locations, without regard to the presence of good restaurants or luxury hotels. Tourists have different social backgrounds, different tastes and different social values: therefore they are attracted to different destinations. This diversity creates opportunities also for those communities so far kept out or marginalized from the tourist market. However, the community has to tackle a series of problems, first of all the lack of international recognition if they do not have any famous and well-known name (Jovanović, 2009).

In the 80^s and 90^s the trend in marketing was mainly to implement advertising campaigns. The logic was to follow a process that began with the creation of the image of the tourist offer, then its distribution on the market and finally its direct promotion to the demand. However, the tourism market is now made up of a more attentive, exacting and evolved demand.

Thus the marketing must be developed in a more comprehensive and complete way, returning to be that for which it was born: a consistent management style. This is especially true in sustainable tourism, where it is essential that the destination marketing governs the business logic of the area according to sustainability criteria, with serious criteria for verifying satisfaction of demand, with constant monitoring of the values of the place, and with continuous corrective actions and organizational development. Therefore, promotion can no longer be the centre of the marketing function. It is the same demand for responsible tourism that does not accept it. In fact it calls for credibility of the value of the advertised information, and for sure it cannot be satisfied with the certifications that are being promoted. The new demand is actually looking for real life experiences. The marketing function therefore must be expressed across the board to produce these experiences on the one hand and to select new targets on the other, making sure in a management way that what offers is consistent to expectations. So there should be much more management and less promotion to improve.

If this managerial function were prevented, certainly more promotional results could be collected, but in addition to spend more and more energy each year to make the promotion credible, neither sustainable development, nor the identity place or much less the support of residents or visitors would be consolidated (Cestari, 2009).

Actually the function of tourism marketing has become an "ethic" tool, which is necessary to prevent the demand. It cannot be directed and governed according to the logic of interest of some people, but in favour of who in the area - residents and visitors - recognize and support the value of the place and its sustainability. Sustainability which does not stand for the original natural environment only, but also for a local cultural tradition and its economic continuity.

Today the function of the tourism marketing can have only one serious purpose: to ensure maximum sustainability to the place. This process is carried out not only through a plan, but also and especially through a tourism management method which consists in producing continuous formulas to "reveal" the value of a place.

As more the revelation of the value is shared by the population and visitors as more it ensures sustainable development to anyone supporting it in practice. It is a form of marketing that have to become necessarily educational, which cannot depend on any biased logic, because the revelation must disclose and promote the most vivid aspects and the closest to the place identity.

A study has been conducted on the influence that the activities of a park or a reserve manager can have on local people or the inhabitants of towns or cities not far from the protected areas, who are often tied to those places because of estate, parentage or simple affection for the most valuable areas next to their residence.

For example, for larger parks there was a strong tourist activity related to short trips with one day visits. This form of attendance of the best places not far from the place where one lives is defined tourism of proximity (Polci and Gambassi, 2004).

Because of their geographical distribution, there is no protected area which cannot be reached in less than an hour from any inhabited place in the countries surrounding the Adriatic Sea. The tourism of proximity plays the role of engine for the tourism promoters in protected areas, and it is a pressure from the bottom that activates the proper procedures for consultations among local actors allowing the start and management of a shared project on tourism. Besides, it is a primary basis for a reasonable economic income, but it is also that wide section of social consensus around the initiative that allows the achievement of the most desired targets through the economic and strategic aid of Local Authorities.

3. The situation in the Mediterranean and Adriatic

In the 70^s we had not yet had a widespread perception of environmental issues. The oil crisis were still seen as a simple economic problems. We had not yet had a general awareness of global phenomena and the sea showed everywhere an amazing ability to regenerate after the first ecological disasters.

It was in Stockholm in 1972, in the first World Summit on the environment, that all countries of the world and, above all, the governments of those countries, began to understand the dimension of environmental issues. At the same time, however, they also got aware of the enormous difficulties that agreeing on the best forms of development to be taken would have met.

The Council of the European Communities in those years reflected on the delicacy of Mediterranean ecosystems. Because of the strong anthropic pressure on its shores, the Mediterranean basin showed a condition of great difficulty under an environmental point of view.

71% of the habitats of Community Importance listed in Annex 1 of Directive 92/43/EEC (Habitat) is located in the Mediterranean biogeographical region. It is no coincidence, then, that the Mediterranean coastal areas are home to the most protected areas in Europe (Parks, Reserves, Oasis or SCI and SPAs, i.e. sites within the European network Natura 2000).

In the Mediterranean, and in particular in the Adriatic, improper fishing techniques have greatly depleted the fish resources over time, and the pollution by rivers, intensive coastal urban sprawl and the installation of polluting industrial plants have further contributed to alter the productivity of a closed sea that, in any case, suffers from its own geographical position and conformation.

A large part of the Mediterranean coasts are undergoing erosion. Coastal aquifers are suffering from a progressive salinization, either due to natural phenomena (limited rainfall, sea level rising, movements of outcropping masses) or caused by human action (reduction of sediment supply to the sea because of containment of rivers or springs, ground water overdraft for industrial and agricultural purposes, overbuilding of land for urbanization, etc.) with negative consequences on the economy and quality of coastal marine waters (Naviglio, 2009).

In 1977 the *Convention for the Protection of the Marine Environment and Coastal Region of the Mediterranean* was approved in Barcelona, with Italy among the promoter countries. From then on this convention had been always called the "Barcelona Convention". It is one of the cornerstones in the international law for the protection of the Mediterranean Sea and an important measure that recognizes an extremely valuable role to the Mediterranean geographical region. Later in this work, this international agreement will be discussed at length. Now it is important to remind that, among other things, the convention acknowledges the fact that there is no sea on the planet where such a combination of unique and universally recognized natural and cultural values has to coexist with an extraordinarily intense and pervasive human pressures as it happens in the Mediterranean.

It would be desirable that humanity was careful in dealing with this issue, in order to find solutions to possible conflicts, thereby ensuring the conservation of the wonders of the Mediterranean Sea. Of course we are working in this direction, but both the commitment and the results are still very limited. Thirty years went by from the announcement of Barcelona, and many steps were taken for the protection of our coasts and sea. The health conditions of the Mediterranean, however, would not seem to have improved yet (Notarbartolo di Sciara, 2008).

It is a very complex issue: there are in force international waters, fishing rights and international codes of navigation. Rivers coming from the hinterland, areas or even countries far away from the coasts flow into the sea. And then the coast: the best places for a development of urban areas because of their temperate climates, ideal spaces for building houses, for the construction of roads and railways and for the development of productive activities.

The protection of the sea and the coast, or even a more improved management of these places, is not easy to implement in the Mediterranean.

Regarding this, there are many measures by the European Union or by individual states and many agreements by regional and international conventions.

One of the most noticeable and recent measures in order of time is the *Marine Strategy Framework Directive*, published in the EU Official Journal in June 2008, which has set targets and deadlines for the Community policy on the subject. The directive n.2008/56/CE is considered a turning point which placed, finally in a systematic way, objective and precise deadlines for the Community policy.[6]

It is a new approach that leaves the typical segmentation of the instruments adopted within the EU, which had always characterized the previous measures for the protection of the marine environment and had not yet led, except in limited aspects, to the achievement of the set goals. This directive has been considered a change of course aimed at establishing a new integrated and innovative policy.[7]

In addition to important previsions for the control of marine pollution, the Directive is to «ensure the conservation and sustainable use of marine biodiversity and to establish a global network of marine protected areas by 2012». Specifically, the planned actions «include spatial protection measures which will help to establish coherent and representative networks of marine protected areas, adequately covering the diversity of ecosystems such as special areas of conservation under the *Habitats* Directive, special protection areas under the *Birds* Directive and marine protected areas as

---

[6] Directive 2008/56/EC of the European Parliament and Council *establishing a framework for Community action in the field of marine environmental policy*. Official Journal of the European Union of 25 June 2008.

[7] Cfr. ROVITO Cristian (2009), *La strategia per la tutela dell'ambiente marino nella Direttiva Europa 2008/56/CE*, Environmental Law, *on line* newspaper. Source: www.dirittoambiente.com (23.12.2009).

agreed by the Community or by the Member States in the context of international or regional agreements of which they are part. [...] At the latest by 2013, the Member States shall make relevant information publicly available, in respect of each marine region or sub-region, on the above referred areas».[8]

The member countries are required to adopt specific conservation measures in a trial programme that, from 2012 to 2014, has its preparation phase and by 2016 «the start of a program of measures oriented to achieve or maintain a good ecological status».

With this directive we are also trying to integrate the Natura 2000 Network in the context of the many network actions to which all the Marine Protected Areas of the world have been moving past time; actions towards the practical implementation of different assumptions on the conservation of an ecosystem as much strongly characterized by the easy mobility of all forms of life as the sea.

A proper course has been taken through this integration to finally overcome what, as has been rightly observed, is perhaps one of the few flaws of the Habitats Directive: «sadly lacking in regard to the Mediterranean marine environment and the species that live there».[9]

The final aim of the Directive is to achieve a good status of the European marine environment by 2021.[10]

---

[8] Parts taken by Articles 1; 2 and 13 of the Directive 2008/56/CE of the European Parliament and Council.

[9] Cfr. ADDIS Daniela (2002), *Attuazione in Italia delle direttive n. 92/43/Cee "Habitat" e no. 79/409/Cee "Uccelli" in relazione alle Aree Marine Protette*, in Community Law and International Exchanges, Quarterly of the European College of Parma, Year XLI No. 3, July-September 2002 Editoriale Scientifica, Naples, p.638.

[10] Cfr. GRONDACCI Marco (2009), Directive 2008/56/EC of European Parliament and Council of June 17, 2008 (...), in the Environmental Law annotated Database (www.amministrativo.it/ambiente, 20.12.2009).

3.1. Fishing

Today, in the era currently dominated by human presence, the so-called Anthropocene, the scientists and the public opinion worry about the increase in the extinction rate of species. The increasing reduction of the number of species has reached such an extent that the current global process of extinction has been defined as the "sixth mass extinction".[11]

About the sea, the situation is even more complex because it is an environment where movements are somewhat simplified for all species. Extinctions at local level, however, are more frequent and, statistically, the 55% of the main threats comes from fishing and a further 37% from the degradation of *habitats*, and, for the remaining 8%, from other factors such as the climate change or the introduction of invasive species. It is clear that fishing is responsible for the significant decreases in fish and shellfish at both regional and local level (Raicevich et al., 2008).

In recent years, the intensification of fishing activities and, above all, the technological development of fishing equipment have generally led to a reduction in fish stocks in the Mediterranean which is now widely accepted.

The exploitation of resources has reached levels of absolute unsustainability, even though it has taken into account the great geographical diversity, both in environments and social realities.

---

[11] Extinction is defined as the moment at which the last individual of a species dies. In the long history of our planet, there were big five "mass extinctions", or events that led to sudden changes in species number and composition. The fifth extinction, the best known, occurred in the Cretaceous, approximately 65 million years ago, and was probably caused by the impact of a meteor on the Earth that put an end to the dominance of reptiles, while mammals prevail in the current stage. Today's period of life of the Earth dominated by the presence of man, for that reason called *Anthropocene*, is leading to a rapid decrease in biodiversity, such as to induce to think of a new episode of remarkable change.

A drastic decrease was recorded, in particular, in cases where the fishing efforts focused on monospecific stocks. In these cases it has been noticed a greater effectiveness of the means of capture due to the widespread use of sophisticated and modern technologies directed to catch almost all of the specimens of a particular species.

Some fishing methods provide for the total capture of the shoals once identified through the use of aircraft in support of the fishing fleets afloat and with the use of sophisticated radar and sonar systems.[12]

If the immediate environmental impact of the most selective fishing affects the commercial stocks of the fish products to be caught, other fishing methods produce even more damage, though indirect, to non-commercial species or to the environment in general. A phenomenon of real destruction of marine ecosystems, for example, occurs on the sandy sea bottoms by the coast where clam trawlers fish by the use of hydraulic dredges. The so-called "turbo-blowers" cause a real devastation of sediments with incalculable harm to the juvenile and sediment dwelling fauna, the main source of food for many fish species of great commercial interest (Vietti and Tunesi, 2007).

But even marine mammals such as whales and dolphins, or reptiles such as turtles, fish and organisms that live in the sea and birds of pelagic species, can be damaged unintentionally or indirectly by fishing gear.

It must be considered that in the Mediterranean for 21 of the approximately eighty species of existing cetaceans are reported. Cetaceans, and sharks are at the top of the marine ecosystem food chain and, therefore, are species of particular importance in order to maintain the balance of nature.

The systems of fisheries that have a greater impact on these species are those not aimed at a very specific kind of fish. Effective but not selective

---

[12] Among the reports, the most recent VISETTI G., *Le reti vuote dell'Adriatico*, newspaper "La Repubblica" of February 25, 2009, insert R2L'Inchiesta.

at all, such fishing systems accidentally catch specimens of species in danger of extinction.

In the Mediterranean Sea it is esteemed a number of accidental killing by fishing gears of about 8,000 cetaceans per year. It is esteemed that every year about 300,000 specimens of cetaceans - 1,000 a day! – in the fishing nets worldwide.[13]

But the real plague is the illegal fishing because it frustrates any common effort directed to make fishing in the Mediterranean a sustainable business. Illegal fishing undermines any conditions of sustainable use of the planet's resources. It damages everybody, at first the fishermen but also the rest of the population, which could soon get aware of the reduction of extremely valuable food resources, either directly affected by fishing or indirectly affected by an imbalance of ecosystems.

The unregulated and unreported fishing is common also in the Adriatic Sea, where it is common practice to fish near the coast and to catch undersized specimens, as well as fishing on sea-grass beds. The extent of the phenomenon and its environmental, economic and social consequences are so serious to be an authentic "priority". In fact, it contributes to the depletion of fish stocks, and often undermines the effectiveness of implemented protective and recovery measures to ensure their conservation. Its weight is likely to cause considerable damage to the economic activities of fishermen and endanger the very survival of coastal communities.

The most discussed tool in terms of illegal fishing is the pelagic drift-net, the famous *Spadara*. It is a net over 2.5 km long, even up to 16 km long, left in the sea to float with the streams in an almost completely

---

[13] Besides this issue, another effect must be considered: the so-called "ghost nets". A real silent massacre, impossible to monitor and control, caused by nets lost or abandoned in the sea which continue to catch fish and cetaceans.

uncontrolled way and, therefore, considered to be more dangerous for the accessory catches than for the target species.

Banned by the EU Commission in 2002 and in the whole Mediterranean since 2005, the *Spadara* is still illegally used. Only in 2005 the Italian Coast Guard seized 800 km of it, followed by 600 km in 2006 (Olivieri, 2009).

A current example of a species that is in danger of disappearing from the Mediterranean is the red tuna. It has been estimated that its capture is three times faster than the reproductive capacity of the present population. Reports have often denounced a hidden market that involves hundreds of vessels equipped with the best technologies for the identification of the shoals and their total catch, including breeding and undersized specimens, all over the Mediterranean. Fattening systems in the waters of other countries allow then to sell the product on eastern markets at very high prices. The red tuna in the Adriatic is now a memory. It is still existing just because this animal is able of high speeds and extended transfers, though it is increasingly difficult to see it.

To control fishery, however, is not easy for a disintegrated and unwilling to control socio-economic context and also for important limitations of the conservational initiatives against such an important business for many coastal communities as we will examine later.

3.2. Topic of supranational importance

An important element that differentiates the protection of marine environment from the world of terrestrial protected areas is that the sea is a collective good as well as all its content of water, life forms and geo-mineral resources.

From the 15th century with the strong presence of the Venetian Republic in the Mediterranean to the Portuguese claims in the Indian Ocean in the 17th, to the most recent unilateral acts aimed at restricting the rights of navigation and use of marine resources to the current recognized Exclusive Economic Zones (EEZ), many attempts have been made to limit the freedom of the seas, but even if constantly moving, international law has always tried to recognize the freedom to carry out in the high sea researches, navigation, collection of materials and fishing as well as the construction of artificial islands and installations.

But if on the one hand there is still a spontaneous and inevitable anarchy in the management of activities at sea, on the other hand, the skills of coastal States to protect the sea environment and life forms are getting stronger and stronger, leading to the question: how to explain the establishment of some Particularly Sensitive Sea Areas (PSSA) on environmental basis in relation with a preordained desire to move trafficking routes to the blue sea or in the EEZ of other countries (Caffi, 2006).

As mentioned above, in the Mediterranean and Adriatic there are factors of pollution and disturbance to the marine and coastal environment coming from the mainland, where the nations involved are not all part of the European Union. It is not obvious, then, that the situation will improve only by working on arrangements in the EU. But, of course, an important step would be if at least the States that took on commitments on this subject implemented them as expected. This in light of the fact that the most industrialized countries are those to create the conditions of excessive use and over-exploitation of natural resources. For that reason the EU countries bear the major responsibility on what is happening in the Mediterranean and Adriatic.

So there are difficulties of different kinds related to the Mediterranean and the Adriatic ecosystems in particular. Seas around different countries and continents have special management complexities.

The process of arranging agreements and implementing procedures is still too long.

In this context it is believed that just protected areas are the best strategic choice, at least in the short-medium term, because, despite the difficulties encountered, they have always guaranteed the maintenance of conservation principles for which they had been identified.

3.3. The selection of marine and coastal protected areas

The policies of protected areas management have long since passed the ancient "development against conservation formulas" with the achievement of the idea that these two factors can integrate each other (Giacomini and Romans, 1982). But it is necessary that the management is carried out through a network, a system of protected areas consistent and representative of all the major *habitats* in the Mediterranean and, similarly, in the Adriatic subsystem in order that protected areas, besides an induced economic development, are a constant warranty over time of biodiversity preservation and conservation, even with an adequate "resilience" of the system in relation to extraordinary events (IUCN-WCPA, 2008).

The data published In the World Conference of IUCN (International Union for Conservation of Nature), held in Barcelona in November 2008, say that Italy is the country with the largest number of Marine Protected Areas and restricted sea surface with 130 km^2 of absolutely protected areas (Abdulla et al., 2008). In addition to these, there are all the protected areas of different denomination, either nature reserves or parks, which are along the coast for want not only of States but also, and above all, of regions and other local authorities.

The existence of management bodies created for these specific marine or coastal areas aloe to address the so-far-mentioned issues as a starting point for a new approach to the territorial government. Associations of local authorities, internal management committees, organizations established for that purpose or other forms of mixed administration in agreement with the associations are all bodies designed for the management of protected areas, in order to overcome the weaknesses of the current administrative formulas by reinventing the socio-economic planning procedures.

However, we also need resources to work in this direction. In any case, in fact, once we find the formula in the institutional activity we would need to activate the economic and financial system that allows to intervene and ensure the protection of environmental resources and the restoration of previously damaged environments.

Economic science has far-back confronted itself with the so-called "indirect costs" which lies on the society for the use of natural resources by any human activity. When these activities are productive, the principle of "polluter pays" is now unanimously recognized (OECD, 1975).

This principle tries to correct the inevitable destruction of some environmental resources in certain production processes through the *internalization* of external costs of exploitation or degradation of resources. The concrete tools, however, elaborated by the economics and operations research, in the analysis of legal forms, seem very inadequate to this purpose. The operation of these tools, in fact, is bound and conditioned by a series of economic variables, also exogenous, that assume not always uniform and predictable trends. In Italy, the fiscal policy instruments introduced on this purpose are not sufficient (Bizzarri, 2004).

As part of the protection and conservation of nature, in recent years the number of protected areas has grown exponentially. The legislature, not only Italian, did well to move in this direction in the light of the above mentioned commitments taken on EU and international tables.

The increasing number of Marine Protected Areas has affected a large part of the sea. However, this process has met many difficulties in recent years.

4. Planning of Marine Protected Areas

While establishing a protected area, the first problem addressed in the collective debate is the definition of its perimeter. A study on this subject is important to properly calibrate the feasibility of planning decisions aimed at a subsequent proper management.

For a whole series of measures relating to different disciplines, the idea of a *clear boundary* has always been very sensitive and controversial.[14] The same delimitation of the whole protected area, considered as a complex and open system, should follow the identification of the included sets, that are the territorial boundaries and functional "systems" acting in the area, avoiding as much as possible to cut the areas of influence and the "field" of relevant and consistent sets, whether they belong to a natural or human order.

In the reality of the territory there are no limits but the linear "zones of tension" expressed as areas tending to mutations. It is therefore difficult to place a boundary line to serve as a clear separation between two different conditions. Moreover, these limitations are aimed at the implementation of the rules of the protected area. These rules cover a wide range of activities, as it is easily deducible. It is quite unlikely that the most logical area of implementation of these rules perfectly coincides for all these different activities. For a delimitation of the protected area, and in particular for its internal zonation, it would be more understandable the use of a *multiple border* made up of various *boundary zones* depending on the activities that they delimit, consisting in territorial zones of appropriate size but variable in the course of time (Giacomini and Romans, 1982).

---

[14] For psychological and social aspects as well as technical and operational ones, this essay is particularly interesting: ZANINI P., *Significati del confine*, Milano, Bruno Mondadori Editori, 2000.

The attempt to move to a planning for *thematic zones* has been carried out in other fields of geography that are not directly related to planning and in some regional landscape plans. Such a system of zoning is characterized by «processes of use and transformation of territorial spaces in relation to aspects of hydrology, biology, history and culture.

In fact, methods of compatible use of a certain territorial portion vary considerably in relation to its environmental features.

A zonal thematic design, in which each disciplinary branch defines not only the size and location of the environmental aspects, but also proposes a zonal sub-level regarding those aspects, or basic rules of use for their specific protection, may be more effective for a more reliable graduation of the possibility of territorial transformation and vice versa, the needs for conservation» (Rolli and Romano, 1995, p.24).

This zonation is related not only to the given geographical area but also to the subject for which the boundary has been set. A thematic zonation could solve out a series of serious problems of the coastal areas, where it is necessary to think of an integrated management of the sea-land zone, which can be more or less large according to the influences mutually caused by the two different environments.

The real implementation of a *thematic zonation*, however, is very complex and involves the formulation of sectorial regulations identifying the specific "vocation" of each place, giving you the chance to better target protection measures. A type of delimitation with this approach could cause many administrative problems. So, just for a matter of simplicity, today we are still acting with linear boundaries, valid for all the themes in legislation, with which both entire protected areas or various internal zone, with different degrees of protection, are limited.

The justification for this choice can be identified only in the temporary nature which must characterize any plan. The structure of any sensitive area, in fact, should be considered as something dynamic that, having a continuous evolution, requires regular alterations and adjustments at its boundaries .

This dynamic or cyclic aspect of plans is certainly the only prerogative in common to all land planning and landscape theories. In case of frequent reshuffles, clear and very-easy-to-locate boundary lines simplify the regulatory apparatus that accompanies each step, and work in the direction of a streamlining for procedures of periodic review.

The need for such plan reviews, moreover required by law in protected areas,[15] does not recommend the use of too complicated zoning techniques, which might hamper the process of plan reformulation to such an extent as to nullify the results of the effort made to differentiate the schemes.

It is an inevitable choice also considering that, as it has been already said: «in favour of this planning approach, beyond any theoretical consideration, there is the mere fact that it really works within the limits of the possible and the overall national situation. And after all this, considering the difficulties with which we can operate in our country, today it does not seem secondary or negligible» (Tassi, 1994 pp.99-107).

4.1. Planning and management of Marine Protected Areas

The establishment of a marine protected area, as well as any other thematic protected area, involves the introduction of restrictions or limitations on the use of environmental resources aimed at the protection and enhancement of natural features and landscapes as well as the identification of new economic opportunities. If conducted on fair basis, this

---

[15] E.g.: the law on protected areas n. 394/91, article 12 paragraph 6 (Plan for the park), says: «the Plan is amended in the same manner required for its approval and is updated with the same procedure at least every ten years».

choice is a joining link of the integration process between the needs for conservation and the needs for development, ensuring an improvement in the quality of life of coastal populations.

In order to respond positively to these multiple objectives, Marine Protected Areas must be properly planned starting from their zoning and an appropriate quantification of the main environmental and anthropic variables within them.[16]

Management and planning forms of Marine Protected Areas thus become the most important tool for the design choices to be made, provided they are aimed at achieving the best levels of efficiency and effectiveness.

The protection and conservation of environments, landscapes, biodiversity and culture of a land makes sense even in a clearly delineated and limited geographical area.[17]

The marine environment, however, is by definition an open environment. The effect of the human presence is often the result of actions undertaken elsewhere: drains from the mainland, disturbances caused by coastal activities, pollution from passing boats, fishing nets dropping from the surface, etc.

The sea is a liquid mass in continuous motion which carries with it soil, materials, life forms and fish. The richness in protected biodiversity within the perimeter of a marine protected area has characteristics of

---

[16] Cfr. TUNESI Leonardo and DIVIACCO Giovanni (1993), *Environmental and socio-economic criteria for establishment of marine coastal parks*, International Journal of Environmental Studies, n.43 -Issue 4- august 1993, London (Uk). pp.253-259.

[17] We are perfectly aware of the limitations of this situation also on land, in particular in the long run, as many studies have highlighted. So we prefer not to go further in our analysis because there are too many differences between marine and terrestrial ecosystems in the connection among ecological systems. Please refer to this bibliography for further information (from Romano, 1996 to Tallone, 2007).

extreme mobility and the sea bottoms, as well as the species that live there, are in continuous and sudden transformation. Coasts are strips of ecosystems with their own features and they are also constantly changing both for natural phenomena and human intervention. Coasts are essential for maintaining the biological balances. The structure of their eco-system is so complex that it has long been thought of specific models to formulate the most appropriate planning forms (Franchini, 1998).

The continuous changes of the environmental context of marine protected areas along with the chronic lack of data and cartographically referenced – "geo-referenced" in technical terms - information[18] are an obstacle of no small importance in the planning process. These are issues on land that can be compared only to the complexity of the planning of active volcanoes' slopes which are ever-changing and rapidly evolving even from a orographic point of view and, moreover, are always within contexts of special protection.[19]

This variability of basis, which amplifies the effects of the developing ecological changes, led to the adoption of important data management information systems, both cartographic and alphanumeric.

---

[18] The available data for marine areas are rarely reported on maps because of the obvious difficulty to turn into graphics an environment extremely complex to measure and draw. This is about the seabed but also and especially about the third dimension, i.e. the content of the above water column. Cfr. TUNESI Leonardo, PICCIONE Maria Elena and AGNESI Sabrina (2002), *Progetto pilota di cartografia bionomica dell'ambiente marino costiero della Liguria*, Quaderni ICRAM n.2, Roma.

[19] Cfr. CAFFO Salvatore et al. (2005), *Il Sistema Informatico Territoriale del Parco dell'Etna, tra gestione del territorio e controllo della qualità ambientale*, showed at the IX National Conference ASITA, 15-18 November 2005, Catania.

In the Central Institute of Research Applied to the Sea[20] various models of work have been developed over the years according to the logic of *Decision Support Systems* (DSS) which could be a valuable support to the management bodies of Marine Protected Areas because they use multi-criteria analytical systems (MCA) combined with geographic information system (GIS) criteria.

Now it is required an analysis on this system that, at national level, seems to be the one that has most attracted the attention of the management bodies of Marine Protected Areas.

GIS, *Geographic Information Systems*, capture, process, store, edit and displays in graphic and alphanumeric form, data of different nature related to a specific territory. These multidisciplinary integrated tools are able to process spatial data, turn them into information, relate different forms of data, analyze and model the phenomena following one another in space and time and, therefore, are able to provide decision support. A GIS is based on the fact that it can study not only "what" (e.g. species list, statistical data, etc.) but also "where" each variable is distributed within the study area. Unlike the classical cartography, which is limited to the reproduction on paper of a single layer of information, GIS contains a series of data which can be correlated one other in every single point and form a part of both planning and management decision-making procedures (Di Nora & Agnesi, 2009).

A DSS which is able to use GIS is a tool that greatly facilitates the comprehension of complex spatial relationships between variables, and can support a participatory decision-making process. It does not replace the decision-maker because it is not designed to prevent the participatory

---

[20] Former-ICRAM, now merged with INFS - *National Wildlife Institute* - and APAT- *Agency for Environmental Protection and Technical Services* - in ISPRA - *Institute for the Environmental Protection and Research.*

process but to get a common synthesis that allows you to view and query databases and information used in decision making. This system can be queried on how a given variable is influenced by the option choice (e.g. the percentage of fishing areas subjected to restriction. The decision maker, or any other user, can query the system to "see" what is protected and what activities will be influenced by applying a specific proposal for zonation.

Once created, the DSS is also an important cognitive reference for the subsequent management of the marine protected area and it goes well with an approach that in recent years has been growing in the world, called "*Adaptive management*", which provides an integrated monitoring in the annual activity management so that, as the name suggests, an "adaptive management" of the protected area occurs (Tunesi, 2009). The interventions or management actions are thus measured and evaluated before and after their implementation and the results are used to refine the subsequent management actions.[21]

This is a systematic process aimed at the continuous improvement of policies and management actions, through the ability to learn from the evaluation of the obtained results. The assessment then becomes a key moment of a circular and virtuous path which is able to self-learn from its own mistakes and successes.

In this context, the method for evaluating the effectiveness of the manager's work becomes an essential element in the process of continuous learning. The tools available to the managers of protected areas for such an assessment are highly diverse, functional, sometimes in relation to the type of natural resource and ecosystem, sometimes to the geographical area. However, the criterion that should guide the best choice among the available tools is the functionality in respect of the evaluation purpose.

---

[21] Cfr. AA.VV.(2007), *Progetto integrato Aree marine protette*, MATTM- Marevivo, Roma. p.51.

Since conservation activities are part of complex contexts, it becomes essential to take into account not only them, but also monitor and embed social, economic, political and cultural variables.

It was carefully studied a way to apply the various methods of management effectiveness evaluation to the Italian Marine Protected Areas. Also through practical experiments, it was proved the importance of having easy and dynamic tools for planning and management of marine protected areas (Franzosini, 2009).[22]

It is said that *adaptive management* is the "last end" for the management of protected areas, while the *management effectiveness* is the "means" to achieve it (Hockings et al., 2006).

4.2. Integrated Coastal Zone Management

The procedure ICZM, *Integrated Coastal Zone Management,* is a form of coastal planning and integrated management provided by a specific Protocol which was signed on January 21, 2008 in Madrid and later became part of the actions provided by the "Barcelona Convention". Compared to the 22 members of the Barcelona Convention, the Protocol was signed in 2010 by 14 countries, including Italy, which was joined later by the European Union.[23] The Protocol came into force on March 24, 2011 after the sixth country among the 15 contracting parties ratified it. Since the

---

[22] This refers in particular to the work of translation, adaptation and implementation carried out on 5 Italian Marine Protected Areas by the Miramare Marine Reserve with Federparchi and WWF Italy and funded by the Ministry of Environment, the manual for the evaluation of the IUCN management effectiveness of marine Protected Areas. See POMEROY R.S., PARKS J.E., WATSON L.M. (2006), *How is your MPA doing?*, IUCN, Gland-Switzerland, Cambridge-UK. The Italian publication that reports the application experience and the translation of this guide is cited in the bibliography: Ministry of the Environment (2007).

[23] Source: www.pap-thecoastcentre.org (30.09.2009).

beginning of its implementation phase, it has become the first normative on a Mediterranean scale.

A strategic approach to coastal zone management must be implemented involving in decision-making all the public bodies which have some responsibility for planning, programming and management of ecosystems.

Italy has encountered some difficulties to start concrete actions in this field despite some initiatives on a large scale, among which the main one called CIP- *Coste Italiane Protette* (Italian Protected Coasts), have been promoted by many since the late 90^s (Moschini, 2006).

Italy has moved in right direction with the participation as a partner in initiatives supported by other countries, such as the "Coastview" project, developed together with 12 European countries between 1998 and 2002[24]. Today this participation is slowly consolidating through other specific projects.

It is worth mentioning that Italy considered the ICZM as one of the most important issues of the "G8 Environment", held in Syracuse in April 2009,[25] and it also reached an agreement with some coastal regions and the Ministry for the Environment and Protection of Land and Sea (MATTM) on a memorandum on the Italian project for the integrated management of coastal zones called "*CAMP Italia*".[26] This project is still in progress under

---

[24] Cfr. NAVIGLIO Lucia, *Strumenti volontari per una più efficace gestione integrata delle aree costiere e delle relazioni terra-mare*, XVIII Festival of the Sea Italy-Tunisia-Malta, MareAmico-MATTM, November 9-12, 2007. Presence in Malta on November 12, 2007, www.mareamico.it (10.12.2009).

[25] Cfr. MATTM, *Linee guida per la sessione III: "Biodiversità, una differente prospettiva*, Siracusa, Environment Minister Meeting, April 23, 2009. www.ansa.it (01.04.2009).

[26] The Coastal Area Management Programme (CAMP) is inserted in the activities undertaken by the Contracting Parties of the Barcelona Convention. The CAMP is

the coordination of the UNEP offices of the *Coastal Management Centre* in Split, Croatia (Naviglio, 2009).

In 2008 the European Union took one of the most important measures on the sea and coastlines in its albeit brief history: the Council Decision on the signing, on behalf of the European Community, of the ICZM Protocol on Integrated Coastal Zone Management in the Mediterranean.[27]

The ICZM, Integrated Coastal Zone Management, begins by noting that coastal areas suffer from everything that happens both upstream in the entire drainage basin and in the sea.

The Mediterranean coasts are about 46,000 km long and look like a hinge that connects two closely related territories. The seabed can be considered as the continuation of land reliefs or, vice versa, the land can be considered a raising of the seabed. The erosion of the mainland, even just for natural phenomena carrying sediments into the sea, with all of their nutrients or pollutants, affect the quality of the marine environment and fishing opportunities. And the "men of the sea", who do not live only on fishing but live in the mainland, are also tied to agricultural activities, urban expansion, tourism, industry.

---

oriented to the implementation of coastal management projects developed in pilot areas in the Mediterranean. Italy was scheduled for an initial study of the characteristics of the territory, operated on a national scale for the identification of a representative group of Italian coastal regions, and a subsequent phase of discussion and consultation with the representatives of the regions selected for the identification of specific areas. To date it appears that five areas potentially suitable for the project have been identified. They are located in the regions of Emilia-Romagna, Lazio, Liguria, Sardinia and Tuscany. Source: www.minambiente.it (02.01.2010).

[27] The EU Council Decision of 4 December 2008 (2009/89/EC) on *the signing by the European Community of the Protocol on Integrated Coastal Zone Management in the Mediterranean*, published in the EU Official Journal on February 4, 2009 (L34/17).

The productivity and usability of the sea depend on how the relations between the anthropogenic pressures and the quality of "maritime resources" are managed, but they also depend on the management of what is happening on land. In addition, the coasts are transitional environments with their own characteristics and peculiarities that simultaneously host a high natural biodiversity and the most of human population. At the same time they are fragile and vulnerable environments, but home to large economic interests related to urban settlements, the creation of ports that facilitate transports, industrial areas and tourist sites. Unfortunately some intended uses are incompatible with others, and the definition of long-term strategies, shared among all public and private bodies to refer in the current management, should precede the implementation of any initiative (Naviglio, 2009).

In short, all interventions aimed at the protection of the marine environment should be included in a set of actions appropriately targeted to the rational and integrated management of a far larger area than the single Marine Protected Area. This is a very well-known issue which today is still not addressed with the right decision (Diviacco, 1999).

There are some methodological proposals for implementing ICZM, but not real procedures. As for the local Agenda 21, actions to be performed have been rather identified through the results of various experiments. The ICZM approach - through the involvement of all the public bodies in charge of land management – should achieve a proper management of beaches and coastal erosion, and the prevention of the effects of global climate change, the management of risks associated with the rise in sea level, the elimination of sources of pollution, the sustainable management of land resources (in the context of agriculture, tourism etc.) and fisheries resources.

The steps to get these results go through an analysis of the existing situation under an environmental, social, economic, cultural and institutional point of view, through the identification of critical issues and assessment of priorities, the sharing of strategic objectives, action and monitoring plans by the use of indicators. These are the same steps that characterize the other voluntary instruments for sustainability (Naviglio, 2009).

5. Programmes, budgets and participation

Interventions in a protected area usually refer to those activities that come along with the establishment of the same protected area and are managed and promoted by the same management body. So, there are interventions that improve environmental conditions and mitigation of disturbing factors for the naturalness of places, scientific research, acquisition and divulgation of results, information for different levels of public, guided tours along terrestrial and marine nature trails, etc.

In this brief list we can already outline three different and closely interlinked types of interventions:
- the first type is the infrastructural improvement of an area with the elimination of disturbing factors;
- the second is about study and research – constant over time and always evolving in the accepted results - useful to get the necessary information for planning activities;
- the third type concerns interventions aimed at developing tourism and environmental education.

The first type of intervention is not a matter of study in this analysis because it does not need any specific planning if not integrated with the implementation of initiatives helping research or tourist development. Apart from special instances, generally infrastructures are not subject of any actions promoted by the authorities of protected areas. Usually, the only thing to do are to demolish the unnecessary and large buildings, remove all those environmental detractors often situated in the marginal areas of the anthropic territory and all those urban furniture and services that have no reason to be in a protected area.

But things are different with the other two forms of interventions which necessarily integrate each other.

Research plays a strategic role also to support the usability management of the MPAs. ICRAM had specific experiences in this context proving the great importance of research to support the management of recreational boating (Agnesi et al., 2006), small-scale fisheries (Tunesi et al., 2004) and underwater fishing (Di Nora et al., 2007; Tunesi et al., 2007). Specifically, the study of underwater diving in the waters of the MPA Portofino has allowed its Authority to collect particularly useful elements. In fact, if properly managed, diving is an activity that contributes to the established goals of MPAs since it is a sustainable tourist activity allowing visitors to see directly the protection effects and extend the tourist season.

MPAs are fully capable of responding to their two primary functions: the conservation of the biodiversity in marine ecosystems and the use of nature as an asset, in a way consistent with the preservation of the ecosystem, by promoting experiences of sustainable development. In this context, research plays the role of a catalyst for a positive circuit which, starting from a proper management of the environment and eco-friendly activities, leads to an increase in "value" of the area under environmental and tourist-cultural point of views, and it can also create a healthy economy related to a "conservative" management of resources (Tunesi, 2009).

Finally, about tourism and environmental education, a protected area can offer various forms of recreational and educational activities so that recreational activities may acquire an educational aspect through the careful study of the results of scientific researches. Visitors should finish their visit of a protected area knowing more than when they arrived, getting useful information for their own cultural growth, because they have learnt everything with fun and even without noticing they were learning.

It is appropriate, however, to distinguish strictly *tourist-recreational* activities from *didactic-educational* ones, because they are organized for different users.

Tourist-recreational programmes are designed for everybody that wants to visit protected areas without any special obligations or needs. The purpose of these activities should not be to concentrate tourists in already crowded places, but to encourage and promote tourism for cultural purposes.

However, the didactic-educational programmes tend to help people understand the significance of nature conservation and to stimulate their respect for it. In this case, *interpreting* and environmental education become effective only if they manage to attract at the same time both cognitive activities, so that young and old visitors can acquire knowledge and understand new concepts, and emotional, with the adoption of new values and behaviors.

In many protected areas it has been created a specific didactic-pedagogical service that provides various educational modules with already-tested lectures, workshops and visits, organized for different durations of stay, from half-day to weeks of study. In this way, the classes have the opportunity to visit museums and visitor centres and enjoy nature trails, workshops, educational areas, aquariums, school watercrafts, etc., always with the assistance of qualified personnel.

5.1. Participation and involvement

A protected area must be based on popular consensus and complete availability in order to have some chances to create an effective planning tool under the complementary aspects of nature conservation and human development.

In fact, it is required that the economic, social, cultural and administrative initiatives that characterize and support it share a common matrix which is as much as possible endogenous. For this reason, many of these systems, that are considered some of the best in territorial planning, resorted to a popular consultation as a basic tool for the evaluation parameters of the performed analyzes. The involvement of local populations is inevitable not to completely miss the target of the plan, and not to raise movements against the same management authority of the protected area.

As well as other protected areas, a marine protected area rests on the consent of the local population and on a proper functioning of all its administrative parts.

Both aspects are closely related because only with the commitment of the authority and local governments in educational and incentive programmes of citizens is possible to create a culture committed to the principles of protection and preservation of a local heritage.

Programmes to be implemented, which are described below in general terms, are indefinite in number. Anything that can help improve the understanding of nature as an asset, of the territory as a resource, in their global importance, is an activity that can be part of what is called here Programme.

There are forms of public participation in planning and implementation of these initiatives. Incentive systems to stimulate the spirit of initiative can be conceived. Forms of co-financing should be identified at the highest levels through programmes of Community interest. A serious programme of environmental education for young people and adults should be implemented. Initiatives should be put in place and events developed not only for fun but above all for the transmission of the cultural message here examined.

The participation of the inhabitants in the planning and administrative activity of the MPA is a point feature of this type of initiative. The ratio protected area-local community assumes a particular relevance and remains, after over one hundred years of debate on protected areas, one of the most controversial and difficult points in the discussion and management of these areas.

The community, or rather the local society, is a complex set of actors who show a certain relative independence and follow their own logics of behaviour. Therefore, in the life of a protected area, participation should be referred to certain roles, which may not be the same for all players.

In addition to the promoters of the protected area and local inhabitants, which certainly take part in the initiative, there are at least three other important subjects: the *local administration* more directly involved in the MPA management, the *other administrations*, involved whenever a project, even of minimal complexity, is to be addressed, and the already existing *local and non-local associations*.

For a correct management of the area and for the implementation of interventions it should be paid attention to the involvement of residents both in the preliminary stages and in the minute definition of the programme. This ensures a correct assessment of the available economic forces, the skills to be used in the direction of sustainable development, the management of the required investments, the necessary training effort and any objection to the project.

To strengthen the sense of local identity and in order to use the protected area as an engine of cultural activities for associations of residents, the population should be involved in the search for themes and contents. This is the first basic educational activity: listening, in which the MPA authority urges the will of citizens and local actors in order to deeply understand – easier said than done - the needs, the will and aspirations of everybody.

A subsequent didactic phase consists in comparison and verification. The collected testimonies are reviewed by the experts of specific subjects (the sociologist, the geologist, the naturalist, the historian, the anthropologist) and are put in relationship with the local environment and with the external reality. The definition of a management form of a protected area, always considered in its broad meaning and not just as a simple nature reserve, is therefore not entrusted to members of the scientific knowledge. The space in which the subjective and emotional view of the individuals is framed (trying not to distort or direct it) in a broader perspective to allow a critical view and an opportunity for the participation of the visitors of the protected area is on the contrary the place to discuss the most heartfelt issues in this human and natural context.

The most ancient and simple way to encourage the participation of citizens, operators, associations and the whole community is to create incentive systems to stimulate economic inventiveness of individuals.

An incentive programme should obviously contain the correct and transparent procedures that the regulations on the use of public resources require. The Regulations for the allocation of resources or the granting of certain benefits must follow the necessary procedure for approval by the executive bodies and the planned forms of advertising.

Within this, however, the activity of incentive to private citizens should be measured in a preliminary analysis that would allow the government to assess the *quantum*, the form and amount of the public action.

Too often in the past very favourable forms of incentive for the private sector triggered that rush to funds which did not guarantee the validity and sustainability of this initiative.

Many activities could be subject to incentives. The only instrument of *environmental reward* included by the legislator in the official regulations of the latest Marine Protected Areas is nothing more than a form of incentive consisting of granting of subsidiary exemptions in exchange for compensation works, rather than giving funds from public to private sector.

Instead, economic incentive programmes of 30-70% of the total expenditure for the improvement of infrastructure or equipment would be more traditional. We are talking about the qualitative renovation of the property, photovoltaic installations, the transformation of polluting enterprises, the replacement of gasoline engines with environmentally friendly forms (electrical, sailing, hydrogen) and many other forms of support to entrepreneurial and associative activities in a protected area.

In fact, the incentive programmes are the best way to help everybody that has no cultural interest in the environmental protection to move towards a more sustainable activity, although the limitations of this instrument are evident when it cannot affect the real intention of its beneficiaries.

PART II: THE SITUATION IN THE ADRIATIC REGION

1. The Adriatic Region and the cooperation

Since the '70ˢ the Adriatic was being referred as «probably the most unified of all regions of the Mediterranean» (Braudel, 1972).

This unity is mainly due to the geography of the basin, given the narrowness of the southern access channel, which is just 72 km wide.

Fig.1.a Adriatic Sea, view from satellite (Source: www.heart.google.com)

A closure that provides unity and control to the whole basin, a peculiarity which was wonderfully used by the Republic of Venice in the entire millennium of its existence.

The history connecting the two shores of the Adriatic is known to all. Perhaps the sea is providing a feeling of separation more today than at the time when the Republic of Venice traded with all the shores without any difficulty. Still in the early 20th century the Adriatic Sea was represented in maps with names of places and cities in Italian, because of the influence that Venice had on the entire Adriatic context, as you can see in Fig.1.b.

However, the hegemony of the important Italian maritime republic had never constituted a political and ethnic unity. The Venetian hegemony is not considered a constant or uncontested condition. Until the end of World War II the Adriatic area was marked by an Italian irredentism and an equally strong counter-irredentism by the Southern Slavs. Even cities like Trieste, under Habsburg rule from 1382, and bitter rival of Venice, had key roles in the maritime political economy that led to the movement of goods, people, languages and cultural influences across the Adriatic and contributed to the unification especially in the northern part.

Fig.1.b Map of the Adriatic Sea in 1906 (Source: www.leg.it/antiqua)

However, the alteration or smashing of the Adriatic area began manifestly in the 18th century and continued until the contrast between Italy and Yugoslavia for the control of Istria. Dismembering the unity of the northern Adriatic, the Yugoslav authorities took the extreme logic of the modern territorial states, trying to tie Istria to far capitals such as Zagreb, Ljubljana and Belgrade rather than to its "unofficial" natural capital: Trieste.

In fact, after such drastic decisions, the Yugoslav Adriatic territories and their resources had never had an important role in the economic strategy of Yugoslavia. The economic orientation of the Yugoslav central authorities consistently preferred the Danube side to the Adriatic one. (Ballinger, 2009).

Further south the experiment by communist Albania destroyed any possibility of cross-border relations, sterilizing the Adriatic border with Italy. It even militarized the Albanian Adriatic coast, filling the beaches of bunkers to oppose a least unlikely Italian invasion.

In the 90^s of the last century the breakdown of Yugoslavia and the fall of communist Albania occurred revealing the miseries, conflicts and poverty that Balkan communisms had concealed beyond the Adriatic. The siege of Dubrovnik, illegal drug and arms trafficking, the crossings of refugees on crumbling motorboats and boats and the disputes over maritime borders are all evidence of the political and economic marginalization of the Adriatic Sea (Gon, 2009).

With the passage of time, however, thanks to the important support of the International Community, with the attainment of independence in Croatia, the entrance of Slovenia in the European Union and the creation of new states such as Montenegro, it newly started the debate on the inclination that the developing process of the Adriatic might be or not be redesigned and revitalized in a future European context.

The Adriatic has traditionally been the scene of "low intensity" movements and exchanges between the two coasts and among the bordering territories on the same coast. The low intensity of these exchanges and their character in some way "physiological", i.e. not compromising the demographic and socio-cultural balance, are the main causes of the lack of social, political and cultural homogenization of the Adriatic area, which still is an area of institutional fragmentation and, at the same time, a horizon for many symbolic references of supranational territorial units (Cocco, 2001).

The Adriatic area has ambivalent characteristics, summarized in the constant combination of unity and diversity in both the environmental and socio-cultural sense. The coexistence of a set of similarities and differences in the various fields provides a specific connotation to this area, an area extremely different in natural, cultural and social terms, which is, however, increasingly being recalled, in a rhetorical way, as a space of cooperation and unification.

The Adriatic Sea is a common territory which is divisible at the same time, where imaginary forms of appropriation - able to recover a wealth of shared symbols and images, but often redistributing memberships according to different tracks and across conflicting geographies - are in action. However, a sharing of institutional trans-Adriatic models is certainly not unthinkable: perhaps today more than ever seems to be at hand (Cocco and Minardi, 2009).

1.1. The Adriatic ecosystem

In environmental terms, related only to natural aspects, the Adriatic ecosystem is very delicate. It is located in the Mediterranean basin which, as we have already said, has serious administrative problems and geographical features that make it a place of extreme complexity.

Some scholars have long thought of the creation of their own Mediterranean sub-network in the eco-Adriatic region.

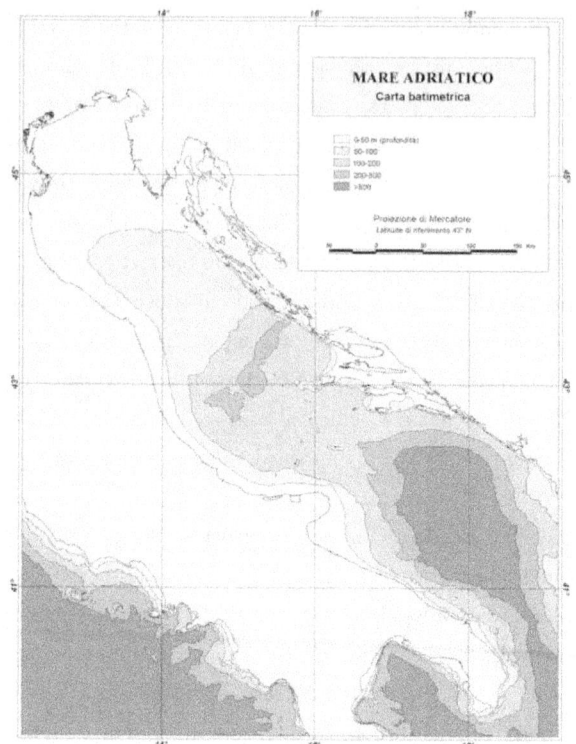

Fig.1.c Bathymetric map (Source: www.izs.it/inadriatico)

A division of the Mediterranean in seven eco-regions was experimentally proposed in the end of the last century. The first published studies reported in the Mediterranean Sea the following eco-regions from west to east: the *Alboran* Sea, Western Mediterranean, Tunisian Plateau/Gulf of Sirte, Ionian Sea, Adriatic Sea, Aegean Sea and Easterm Sea (Spalding et al., 2007).

The eco-region is a large land or water unit containing a geographically distinct combination of species, communities, and environmental conditions. The limits of an eco-region include an area wherein important ecological and evolutionary processes interact with a lot of force.

The eco-regional conservation is an evolution in thinking, planning and acting with the most appropriate spatial and temporal scales for a full success of the conservation of biodiversity (WWF, 2003).

The publication of the EU Marine Strategy Framework Directive has recently made official the existence of a geographical area defined as "Adriatic Region".

Articles 3 and 4 of the Directive state:

«Article 3. Definitions
For the purposes of this Directive, the following definitions apply:
.... omission
2) Marine Region: Region referred to in Article 4. The marine regions or subregions are designated to facilitate the implementation of this Directive and are determined taking into account hydrological, oceanographic and biogeographic factors;
.... omission
9) regional cooperation: cooperation and coordination of activities among Member States and, where applicable, third countries that are part of the same region or subregion, for the development and implementation of strategies for the marine environment;
Article 4. Marine regions or subregions
1.Member States shall, in fulfilling their obligations under this Directive, take into account the fact that marine waters under their sovereignty or jurisdiction form an integral part of the following marine regions:
a) Baltic Sea;
b) North East Atlantic Ocean;
c) Mediterranean Sea;
d) Black Sea. ... omissis
In order to take into account the specificities of a particular area, the Member States may implement this Directive based on subdivisions, at the appropriate level, of the marine waters referred to in paragraph 1, provided that such subdivisions are delimited in a way compatible with the following marine sub-regions:
... omission

b) in the Mediterranean Sea:
i) in the West Mediterranean Sea;
ii) the Adriatic Sea; ... omission» [28]

The Adriatic Sea region is thus officially recognized as a marine environment to protect.

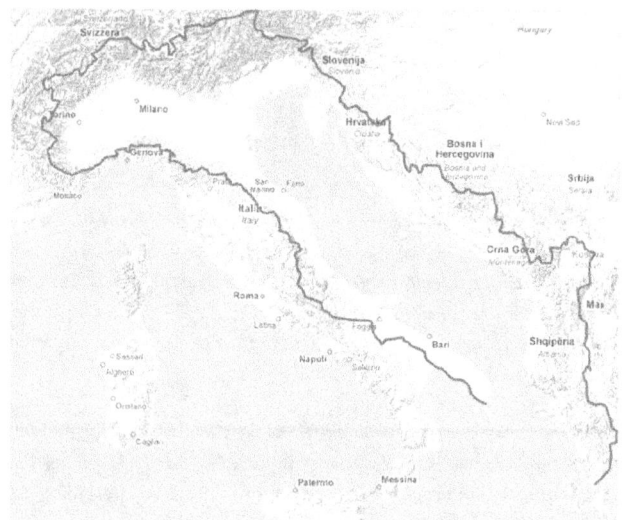

Fig.1.d Map of the Adriatic drainage basin (Processed by: www.googlemap.com)

But when it comes to the marine environment, we cannot help thinking of the waters poured into the sea by rivers and we cannot help considering that some of the most extensive river basins recognized in Europe affect just the Adriatic Sea.

---

[28] Directive 2008/56/EC of the European Parliament and Council of 17 June 2008 establishing a framework for a Community action in the field of marine environmental policy (*Marine Strategy Framework Directive*). Official Journal of the European Union L 164/19 of 25.6.2008..

From a cartography for the identification of basins of European interest, it is clear that the following river basins of European interest affect the Adriatic: River Po and Adige River (Italy), *Neretva* River (Croatia and Bosnia-Herzegovina) and *Drin* River (Albania).

In this context, a hypothetical perimeter of the Adriatic region could also be considered the drainage basin of competence, coinciding with the boundary shown in the map in Fig.1.d.

The Adriatic basin has important environmental, social and economic issues. Despite the importance of these issues, we have the feeling that central institutions only occasionally deal with them, especially after the Authority for the Adriatic was abolished, with a passage of a few lines in an Italian law of 1993.[29]

It has been noticed for some time as sea protection is a public interest that must unfortunately deal with other powerful interests more or less directly connected with the business and the market system. In this confrontation in the management of the situation in the Adriatic, a large number of ministries, agencies and interests are involved, but rarely environmental issues are important for them (Di Plinio, 1994).

1.2. The economic situation in the Adriatic countries

The 90^s and the first decade of the millennium seem to have been years of conflict in which the will to make Europe a single large "home" where accommodating the needs of all the people living in it apparently did not match well with the re-emergence of particularism, local biases, nationalism and religious feuds. In recent years the Adriatic has been the most interesting "challenge" for sustainable political, economic and social

---

[29] The Law of 24 December 1993 537, *Financial corrective actions*, in Article 1, paragraph 30, states: «*The Authority for the Adriatic is abolished and its functions are transferred to the state administrations responsible for the matter*».

development of the new millennium, representing a historical region, declined to semi-periphery in an economic world in search of a new identity in the international context, an identity to enable it to play a leading role in the emerging European Union, trying to move its focal point more southward (Cardinale, 2006).

An analysis of the economic situation in the Adriatic countries is not simple because only three countries - Italy, Slovenia and Greece - are members of the European Union out of 7 (or 8 if we want to consider also Serbia), including Albania, Bosnia-Herzegovina, Croatia and Montenegro.

This creates an obvious lack of homogeneity of the data collected by statistical institutes or research centres that often do not use the same survey systems. So, to avoid to use heterogeneous data, in a summary examination of the economic situation in the Adriatic countries, it was decided to adopt only very general parameters provided as unitary from sources such as the World Bank or the International Monetary Fund, without going into too much detail in order to avoid easy interpretation errors in comparing indices which, by construction, have a different information content. The difficulties found in the availability and stability of the existing databases must be added to the great difficulty in reading the current economic picture.

The dynamism of the economies of the eastern Adriatic countries, due to the presence of young and newly established states (such as Montenegro), increases the complexity of the internationalization processes. Moreover, in absolute terms, the current reality is marked by an unprecedented economic crisis that has devastated the international context, throwing it to an extremely volatile dimension which is difficult to interpret.

Fig.1.e Political geography map (Processed by: www.d-maps.com)

For these reasons, in the following description of the economies of the single countries, apart from Italy and Greece, considering the difficulty in interpreting the current economic context, we have tried to give a picture of the current situation referring to predominantly journalistic sources. The collected data come thus from calculations made from the newspaper EcoAdriaNews of Ancona on the basis of updated data to 2010 from: International Monetary Fund; INSTAT; ICE, the Bank of Albania; Statistics Agency of Bosnia and Herzegovina, Croatian Institute of Statistics, National Bank of Croatia, Greek Ministry of Economy; Slovenian statistical Office.

The picture coming out on the basis of this information is as follows.

Albania

Now Albania is an institutionally stable country. It has started a process of internal reforms aimed at bringing its institutional, administrative and legal system to western standards. The GDP growth of the country, rising for nearly a decade now at a rate of more than 6% per annum, suffered a decline in 2009 due to the international crisis, remaining thereafter, according to the latest IMF estimates, at levels which amounted around 1.5%. The adverse effects of the global crisis in the real economy occurred in a significant reduction in remittances from Albanian migrants (reduced from 18% to 15% of national GDP), in an apparent slowdown in what remains the leading sector of the economy, building, and in a decrease of trade with foreign countries (exports in January 2009 decreased by 11.6% over the previous year, while imports, also in January, fell by 10.5%).

Albania remains one of the poorest countries in Europe, with a significant percentage of the population still living below the poverty line (18.5%), although official Albanian sources reported a reduction of extreme poverty of 3% in 2008. About 60% of the workforce is employed in the agricultural sector which occupies 20.6% of GDP, compared to 19.9% of the workforce employed in industry and 59.5% in services.

The progress made by Albania in *governance* and in the creation of a favorable *business climate* have been appreciated by the World Bank and International Monetary Fund. In 2008 a high flow of trade with foreign countries was confirmed, always characterized by strong imports and a weak export trend. Its rate of coverage remains low (25.6%).

In 2012, the GDP per capita of Albania was $ 8,000.

Bosnia Herzegovina

From an economic perspective, Bosnia and Herzegovina is a country in constant growth since 1995, engaged in the transition to a (mixed) market economy fully self-sustainable (international support is still significant). The fundamental economic data of the country, despite the world financial and economic crisis and its impact on the Bosnian economy warrant cautious optimism.

According to World Bank estimates, the GDP that in 2008 recorded EUR 11.25 billion (+6% over the previous year), with GDP per capita of 2,960.11 euro, in 2009 had a further small increase (+1.5%). There were many projects of the Community programme IPA (Instrument of Pre-Accession) for Bosnia and Herzegovina for the period 2008-2010, for a total amount of 269.9 million euro. 254.5 million euro, the largest part, should be earmarked for projects to support transition and institutional building in the country. Cross-border cooperation activities will be financed with the remaining € 15.5 million.

With regard to the World Bank, the loan portfolio of the new "Country Partnership Strategy 2008 – 2011" for Bosnia and Herzegovina was 200 million dollars. The strategy identifies priority areas such as infrastructures, support for investment, government spending and services.

In 2012, the GDP per capita of Bosnia Herzegovina was $ 8,300.

Slovenia

The only country of the former Yugoslavia already in the EU, Slovenia is experiencing all the effects of the international economic crisis and its exports have had a sharp decline in recent years.

Unemployment rate increased up to 7.8% in December 2008, but the figure is rising. It was significant also the decline in exports of goods (-10.2%) and imports (-11.5%). In a difficult general situation, Slovenia, according to all the major credit rating agencies, is ranked among the most reliable countries in Central Europe. It has a high degree of openness to international trade and foreign investment, with a prevalence of exports in particular in the sectors of non-ferrous metals, oil and fuel, equipment and electrical machineries, engineering and components, electronic products and components, textiles and clothing, wood and furniture manufacturing, plastic and rubber products, motor vehicles and parts.

It imports, instead, machineries, metals and metal products, non-ferrous metals, motor vehicles and parts, chemical and pharmaceutical products, textiles and clothing, electronic products and components, agricultural and food products.

In 2008, Slovenian imports exceeded 23,000 million euro, compared to just under 20 billion of exports. Germany, Italy, Austria, Croatia and France are the best trading partner countries.

Proclaimed by the National Geographic the fifth destination in the world for attractions regarding environmental and land protection, preservation of artistic and cultural heritage, Slovenia has developed in a few years the primacy of green oasis for what concerns the environmentally sustainable tourism. On the phenomenon of eco-tourism, globally increasing, the Slovenian government is aiming at the development of rural areas and the rediscovery of pristine territories, boasting a surface covered for 60% by woodlands, ideal for those who love outdoor tourism and mountains, always with an eye for responsible tourism and ecology. The quality of coastal and marine areas is not less important, though they are still in a transition phase for the difficulties in the resolution of the limits of territorial waters with Croatia. In 2012 the GDP per capita of Slovenia was $ 28,600.

Croatia

Croatia is a country that suffers from some macroeconomic imbalances which threaten its solidity, such as the high proportion of foreign debt and the deficit in the trade balance. In this context, since 2007 the Central Bank has started a restrictive monetary policy that has made possible to address the international crisis with a sound financial system and to maintain a stable exchange rate of kuna against euro. The rating of this country has remained stable and the feedback of international agencies reflect a moderate risk. More than 90% of Croatia's trade with the rest of the world is now governed by the principles of free trade or concessional terms.

The value of trade interchanges is growing at very lively levels and at the end of 2008 reached 30.4 billion euro (+9.2% compared to 2007). Export sales were of 9.6 billion euro (+6.4% compared to 2007), while imports exceeded 20 billion euro (+10.5%). Over half of the trade is carried out with five countries, of which three are EU members - Italy, Germany and Slovenia -, plus Russia and Bosnia and Herzegovina.

Croatia is an ascertained outpost of the new Balkan tourism, but it is surprising the persistence of this data, proving every year the Croatian State to be one of the most popular destinations. The data confirm a positive trend, with an average growth of 5% for the tourist presence in the County of Istria, 6% in the County of Sibenik, 10% in the County of Dubrovnik-Neretva, up to peaks of 14% in the counties of Split, Dalmatia, Lika and Senji. In 2012, the GDP per capita of Croatia was $ 18,100. In 2013 Croatia joined the European Union.

Montenegro

Montenegro's GDP grew by 8.6% in 2006 and 10.7% in 2007 and 8.1% in 2008, according to data from the Montenegro Institute of Statistics (MONSTAT), and it has been in line with this rate in recent years up to a total of 2.5 billion euro.

In recent years the macro-economic parameters of the country have been in general characterized by a steady growth of GDP, an oscillating inflation, the strong growth of trade with foreign countries, but also by a large deficit in trade (the largest in the area) and a wage growth exceeding that of productivity. Inflation reached 7.8% in 2009.

The keys to the growth of the country are the tourism sector and the revenue from foreign direct investment (FDI), also related to tourism, but also infrastructure, finance and energy.

The composition of the gross domestic product of Montenegro is dominated by the service sector (56.2%), followed by manufacturing (11%), agriculture (6.9%) and construction (3.4%). Since 2008, the Montenegrin exports have declined significantly, further exacerbating the large deficit. Even the real estate industry, one of the engines of development, is experiencing a sharp decline, with many planned investments that have been postponed or canceled. Montenegro adopted Euro as national currency. A prudent and restrictive fiscal policy is the only lever control of macroeconomic problems. The IMF, in a recent analysis, predicted a slowdown in GDP growth in the coming years (about 5%), to be reviewed because of the financial crisis of the world market and the strong orientation of the country towards tourism and FDI. In 2012, the GDP per capita of Montenegro was $ 11,700.

1.3. Cross-border cooperation

Since World War II, in Europe the perception of borders has changed from lines of definition and separation to development and cohesion areas. The change in the perception of borders has been going at the same pace as the project of European unification. It has even exceeded the narrow borders of the Union involving different areas of Europe. In this direction the commitment of the Council of Europe[30] for a stronger peace and integration

---

[30] The Council of Europe should not be confused with the European Council, or the regular meeting of the Heads of State and Government of the Member States of the European Union. The Council of Europe, established on May 5, 1949 with the Treaty of London, brings together 47 countries inside and outside the EU and its purpose is to encourage the creation of a common democratic and legal organized

process in all Europe and an increasing attention-seeking of local authorities in the process of European integration have played a leading role (Coletti, 2009).

The European Commission has worked extensively on the issue of international cooperation and introduced the concept of decentralized cooperation intending it as a new form of development based on the principles of environmental sustainability, participation and human development. The Commission's definition does not attach particular importance to local governments, which are placed on the same footing as other parties, different from central governments, which promote or are carriers of participatory development, such as non-governmental organizations, trade unions, churches and religious organizations.[31]

The European Commission's definition is at the opposite end of the procedure adopted for example by France which, for its own history and extension of its business beyond the State borders, is considered one of the countries with the longest tradition in the field, where the decentralized cooperation is the cooperation of local governments and concerns almost exclusively institutional and public administrations.

The Italian way to decentralized cooperation, as it has developed to date and how it was later defined by the Italian Foreign Ministry, is an intermediate hypothesis, in which the role of the institutional entity in a given territory is considered prevalent but not exclusive. So the decentralized cooperation is the action of development cooperation carried out by local authorities in partnership with counterpart institutions in other

---

area in accordance with the European Convention on Human Rights and other reference works on the protection of the individual.

[31] Regulation (EC) n.1659/98 of the Council of 17 July 1998, OJ of 30 July 1998 on decentralized cooperation.

neighbouring or cross-border countries or at least countries with which cooperation processes have been started.

The added value of decentralized cooperation compared to traditional governmental cooperation and non-governmental cooperation can be identified in different sizes.

- First, the commitment of a local self-government in the international cooperation and towards another territory has a greater political value than the action of a single NGO, as it ideally represents the commitment of an entire community in favour of another community.
- Second, the local governments can mobilize all the resources and expertise of the various actors in the territory, filling with contents the partnership with other local areas even in case of limited financial resources.
- Third, the decentralized cooperation often involves support to the processes of decentralization and local development of the partner areas, which is in some ways a natural consequence of cooperation activities also on specialized topics (Rotta, 2009).

About the decentralized cooperation that has characterized the experience of Italian regions, provinces and municipalities – of which there is a wide inventory with more or less positive and structured experiences - the territorial partnership is both an ideal model, which explains the principles and guidelines of international cooperation, and a long-term goal to achieve.

The Partnership Action is at once a natural evolution of the decentralized cooperation models so far practiced by the Italian local self-government, a methodology to create stronger relationships with territorial partners, and a point of arrival of cooperation activities.

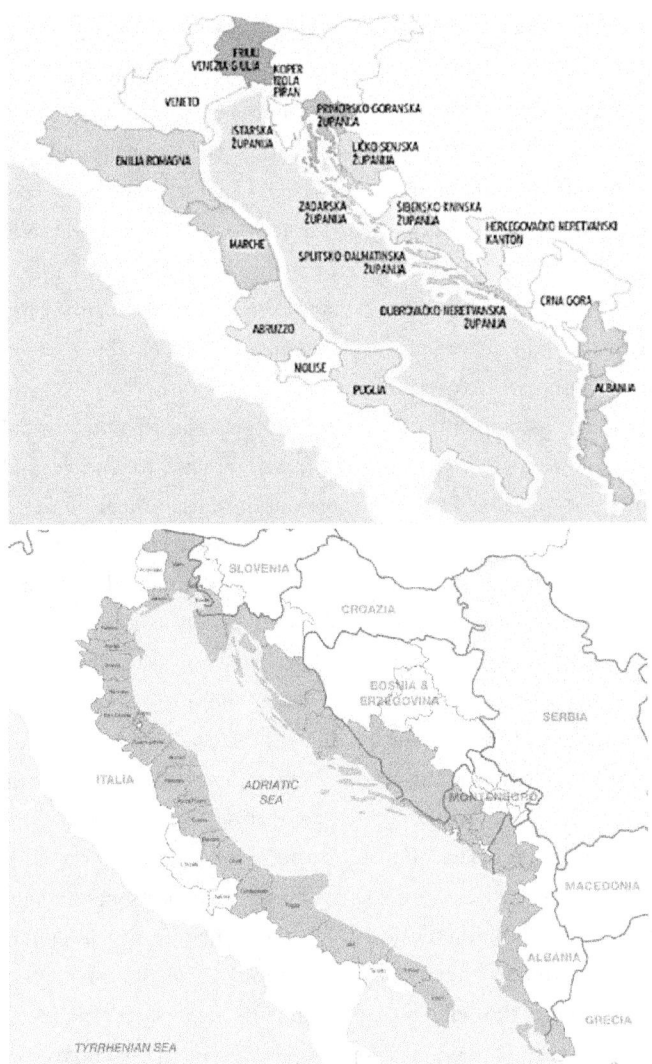

*Fig.1.f Political geography map: Regions and Provinces
(Processed by: www.d-maps.com)*

A better focus on the concept of territorial partnership could have practical advantages for the action of local self-governments, as it would allow to identify and systematize best practices in international interregional cooperation, and to support with more strength and contents the role of regions, local self-governments and territories in the multi-level *governance*, definition and implementation of external policies of pre-accession and proximity to the European Union (Stocchiero, 2004).

Here the term "*governance*" stands for the involvement of parties out of the institutions in charge of territorial government on the implementation of public policies and in government planning.

It is an assignment of political responsibilities and administrative skills, once the prerogative of the central government, to a wider audience of semi-public or private actors (horizontal *governance*) and local or supranational ones (vertical *governance*). In one word: "governing without government". The term "multi-level *governance*" highlights especially the vertical side of the redistribution of power and the knowledge that any policy, regardless of the priority level to which it is formally attributed, necessarily requires the interaction among a plurality of entities acting on distinct geographical and institutional levels (Scarpelli, 2009).

Territorial partnerships mean cooperation basically as a support to development processes, rather than generating development through the development and implementation of projects. It recalls thus the shift from a project-based approach (led by the offer and by experts in the short term) to strategies and programmes (led by the demand, which enhance local resources, characterized by procedural nature and medium-long term). The agreements among the sub-state governments express consequently multi-annual programmes of joint development, based on the comparison of their policies that may provide support to the public budget of the local authority partner as well as a sequential and flexible series of cooperation measures (Rhi Sausi et al., 2004).

The Italian decentralized cooperation was born and largely developed in response to the conflicts in the former Yugoslavia in the 90^s and, to a lesser extent, to the difficult situation in Albania. Many local governments debuted in the international arena by financing or supporting initiatives of non-governmental organizations or individual citizens of their territories, entering, in this way, in direct relation to the territories of former Yugoslavia and with their local authorities. The Balkans were a sort of training in which the actors of Italian decentralized cooperation were formed, from early interventions - uncoordinated, episodic and related to the humanitarian emergency phase - to a more conscious and structured approach, developing a culture of cooperation and peace, devoting increasing human and financial resources to co-operation activities and tending to the partnership model (Rotta, 2009).

The Adriatic space as an interdependent system, rather than a barrier, is an opportunity for development that has in the territorial partnerships a specific variation and a concrete tool. About the different dimensions of development, partnerships among territories can transform the Adriatic from a boundary line into a point of transition among socio-economic systems at different levels of development, in a highly integrated and close system.

The interest in cooperation is naturally diversified depending on the geographical, political and national position of the single territories, so the more developed regions tend to be more interested in safety, the economically disadvantaged territories in prospects of development. Thus, there is a sort of potential trade-off between security and development, which may bring together the interests of both parties towards a partnership.

The management of common goods, such as environmental heritage and fisheries resources, evidently refers to the need for a cooperative approach between the two sides of the Adriatic (Rotta, 2009).

1.4. The Adriatic-Ionian Initiative

Currently the Adriatic-Ionian Initiative (AII) is bringing forward the most interesting initiative for the coordination of activities in the Adriatic region.

It is an initiative launched by a Conference on Development and Security in the Adriatic and Ionian Sea, held in Ancona on 19-20th May 2000, which was attended by the Heads of Government and Foreign Ministers of the six coastal countries (Albania, Bosnia-Herzegovina, Croatia, Greece, Italy and Slovenia).

At the end of the Conference, the Ministers of Foreign Affairs, in the presence of the European Commission, signed the "Declaration of Ancona", stating the importance of regional cooperation as a means of promoting economic and political stability, necessary conditions for the European integration process. In 2002 the union of Serbia and Montenegro joined the six original members. After the split of the federation in 2006, both countries kept their *membership* in the Initiative, which is currently made up of eight countries.

The establishment of the IAI was aimed at strengthening the regional cooperation between the two shores of the Adriatic in order to promote shared solutions to common problems, especially related to security and stability in the region but also to the environmental protection of the Adriatic and the Ionian basin.

Ten years later, the geopolitical context of the Adriatic Ionian Initiative changed deeply. In particular, Slovenia became a member of the European Union in 2004, and the other countries of the eastern AII (Albania, Bosnia and Herzegovina, Croatia, Montenegro and Serbia), albeit with different times and ways, started a process of approach to the Community institutions according to the Stabilization and Association Process and in view of a final integration into the EU. However, the reasons that led to the establishment of the AII have maintained and even increased their validity over the years. Due to the increased interdependence among the States inherent in the processes of globalization, the joint solution of the problems affecting the Adriatic region requires an additional level of cooperation, not only between the countries of the region but also among regional initiatives. The cooperation has therefore taken on new forms, such as partnerships among local actors.

The decision-making body of the Adriatic Ionian Initiative is the Council of Foreign Ministers (Adriatic-Ionian Council), whose agenda is processed in the course of regular meetings among *Senior Officials*, which are held three times a year. The Presidency rotates annually according to an alphabetical criterion and the turnover generally occurs between May and June. Italy succeeded Greece on 1st June 2009 and since May 2010, the task has been taken by Montenegro.

In June 2008, thanks to the support of the Marche Region, a Permanent Secretariat of the Initiative was opened in Ancona.

The regional proposal to establish a Secretariat for the Adriatic Sea with the task of carrying out political action and support for multilateral relations, to encourage the use of existing opportunities at a national and Community level and to give a permanent and certain home to relations and contact to public and private entities that operate in the area, represents a concrete attempt of coordination in continuity with the Adriatic-Ionian Initiative and the Charter of Ancona promoted by the Ministry of Foreign Affairs. It is interesting to note that this coordination is situated "downstream" of the various existing national and Community tools, and it partially compensates their "upstream" deficit of coordination (Ianni and Toigo, 2002).

The purpose of the Secretariat is to ensure continuity in the transition between the two Presidencies and give a *"project oriented"* slant to the initiative by operating as a catalyst for the proposals by the member countries.

The AII Permanent Secretariat has started a cooperation with the Adriatic-Ionic *fora* already active in the region: the Forum of the Chambers of Commerce and the Forum of the Adriatic and Ionian City and UniAdrion, and recently with the network of Adriatic protected areas AdriaPAN.

During the Italian AII Presidency of 2009-2010, it strongly emerged among the member countries a shared interest in enhancing the Adriatic-Ionian basin and its different forms of territorial cooperation through an integrated strategy to support the completion of its European integration and to promote sustainable development, bringing back the plurality of actors and initiatives operating in the region into a common framework.

The European Territorial Cooperation, already target of the Cohesion Policy of the European Union, has developed into many forms and initiatives in the region, and this plurality of interventions, operated by the States at central and decentralized levels, as well as by non-governmental associations and cross-border representatives of the civil society in the coastal countries, requires coordination and systematization to better achieve the objectives of social, economic and political development towards the creation of an Adriatic-Ionian macro-region, in terms of the European Union.

The reference to a macro-territorial entity in geographic terms, including all political and institutional implications, if it has further corroborated the process of EU enlargement, has also given a positive endorsement to all cooperative activities which have been undertaken so far in various sectors by local, regional and national institutions which are the reference point for networks of companies, associations, universities, protected areas and other non-profit bodies gathered in several cross-border programmes that the European Commission has put in place in recent years.

The structural components of the relations between the two shores of the Adriatic are therefore reinforcing. So, if the EU policy seems to have already found its own path of reference through enlargement initiatives, with the conclusion of negotiations for association agreements and then the full membership of the European Union, also national policies must rapidly reconfigure themselves in terms of economic, social and cultural rights to adopt strategies and tools and give continuity, regularity and therefore legitimacy to the many initiatives that are being launched and developing in different sectors of economic and social life of the Adriatic peoples.

At the present time a fundamental purpose for the creation of a common Adriatic identity is standing out with growing evidence: to preserve, promote and disseminate information and tips about the environment, traditions, economy and culture of the Adriatic-Ionian region to be considered as a European Border Region. A cross-border territory that is to be gradually integrated into the European Union as *Macro-Region*, which includes different social and territorial situations including: cities, ports, islands, religious and monastic communities, coastal towns, inland villages, mountains, lakes, rivers and protected areas (Minardi, 2009).

2. The Adriatic protected areas

To define a framework of Adriatic protected areas is not simple for both the complexity of the choice about the geographic area to refer and for the type of protected area to be taken into account.

Our purpose is to identify how protected areas can create opportunities of development in the Adriatic area. In this context, the protected areas to be taken into account are those belonging to countries that affect the Adriatic area up to its most southern limits, which geographically can be considered Ionian, and are directly related to the marine ecosystem, such as Italy, Slovenia, Croatia, Bosnia-Herzegovina, Montenegro, Albania and Greece.

It was decided to consider under a geographical point of view all the marine protected areas in the Adriatic Sea and in the northern part of the Ionian Sea and also all those other protected areas on land which border the sea in at least one point in their perimeter or, according to the laws of the various countries, in what is considered the boundary of coastal state property.

This option makes the research study the Adriatic basin only about the marine part and not the whole drainage area, which, as we have seen above, could be a viable option, and perhaps also the most correct to follow if the environmental aspects of the region were mainly considered. This choice limits considerably the number of the protected areas to be examined, at sea and along the coasts of the seven countries of reference, providing a more homogeneous data and information collection either from the physical-ecological point of view or the geographically-economic one.

In fact, considering all protected areas affecting the drainage basin would have extended the examination to an extreme number of protected areas, including much of the protected Apennine and Alpine areas (e.g. the Gran Paradiso National Park in the Aosta Valley), and inevitably changed and falsified any idea of development and protection for areas closer to the Adriatic marine environment.[32] It was decided to leave out from this study

---

[32] Including the geographical area resulting from river basins, which would be a proper way to approach the subject from an ecological point of view, would have led

protected areas that did not have a direct contact with the marine and coastal environment for the big difference among their habitats, ecosystems and various forms of economic development.

A further selection had to be made because many of the so far considered Adriatic marine and coastal protected areas are punctual elements of extreme natural and landscape interest, such as a rock on the high sea or a monumental tree on the coast, with no reason to entrust any specific management to a public or private authority in order to develop forms of appreciation and utilization of the resource either in economic terms or for the protection of the good. The areas that it was decided to exclude are small areas situated in all the countries wet by the Adriatic Sea. They are called "Monument" or "biotope" or "Individual Good", etc., for which only forms of passive protection are provided through stringent rules of protection that do not imply any form of management leading to a sustainable development.

Once outlined the basic criteria for the choice of marine protected areas, the research was performed starting from the largest database available on the international scene to have a list as large as possible of Adriatic protected areas. From this list, we proceeded to carry out a selection of those protected areas responding to the characteristics listed above. This selection was performed starting from both cartographic and

---

to consider as Adriatic protected areas Parks such as the above-mentioned Gran Paradiso in the Aosta Valley, but also many other vast National Parks such as Stelvio, Val Grande, Tuscan-Emilian Apennines, Foreste Casentinesi, Sibillini, Gran Sasso and Monti della Laga, National Park of Abruzzo, Lazio and Molise, even the Alta Murgia, adding also masses of regional Parks and Nature Reserves that, compared to the economic value of the small Adriatic coastal and marine protected areas, would outline a completely different framework for their internal management systems as well as for their socio-economic characteristics related to completely different forms of tourism.

bibliographic materials available in physical and virtual archives and libraries, and then making a check through a direct survey on a large number of stakeholders which were conveniently selected on the basis of numerous sources of information.

2.1. Bibliographic and cartographic research

The use of computer search systems on the web and the activity of bibliographical research has led to consult various sources, and the one that was most used for its richness in both cartographic and alphanumeric information is "Protectplanet.net".

It is a site built thanks to a collaboration among UNEP[33], the UN Environment Programme, IUCN[34] – the International Union for Conservation of Nature; CBD[35] – the Convention on Biological Diversity and WCPA[36] - the World Commission for Protected Areas, under the two projects WDPA - *World database on Protected areas*, the worldwide database on protected areas and MPAGlobal - *Marine protected Areas in Global* - which is the application on *Google Earth* of the WDPA for marine protected areas.

"Protectedplanet.net" is an online interface based on the new *Social Networks*, which provides detailed information on the protected areas of the planet.

---

[33] United Nations Environment Program.
[34] International Union for Conservation of Nature.
[35] Convention on Biological Diversity.
[36] World Commission for Protected Areas.

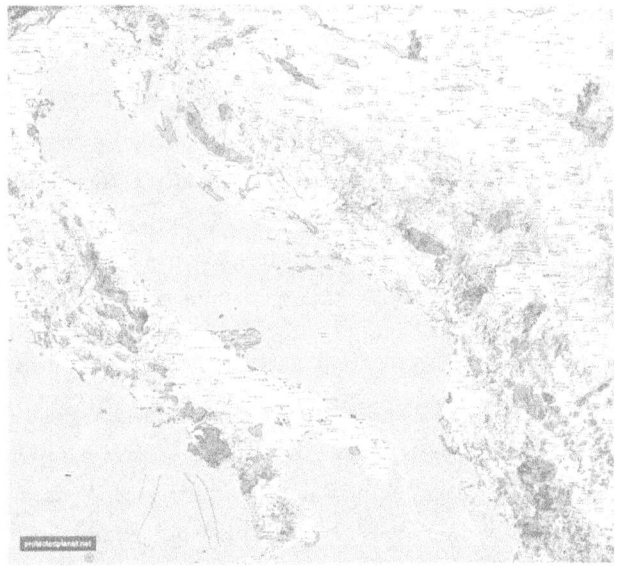

*Fig.2.a Map: Natural protected areas in the Adriatic region
(Source: www.protectplanet.net)*

Using the latest satellite images "Protectedplanet.net" allows users to quickly identify different protected areas such as national parks or marine reserves in the world. You can get accurate and detailed information on threatened species, types of vegetation and their living conditions, or even on the resources in that specific area or ecosystem.

The real innovation of this site is that "Protectedplanet.net" also offers visitors the chance to become "authors" of the Site and to provide information on the selected places, upload/download photos of trips in protected areas, write travel stories, or recommend places of interest, to share later on social-networking sites such as *Facebook*, *Twitter*, *Flickr* or others. [37]

To allow the user to interact with the consulted database and even add information is a system more and more used also in the world of scientific research. At the beginning, the possibility to modify information on the database was allowed only on websites for amusement and entertainment. But, with the passage of time, experiences such as *Wikipedia*, the encyclopedic site with the world largest number of information strings, brought also the scientific world to open out the possibility to acquire information.

So today there are many sites linked to important scientific entities that, in order to get information, enables users to enter data on a certain string. Usually, the entered information, as in the case of *Wikipedia*, are first checked by an inspection body, mostly to avoid crimes against public decency and good morals, and not so much to control the substance of the information, and then entered in the network.

The principle at the basis of this method of information collection is the control and verification by the community. The maximum access to all information for all users who connect to the network proves the effectiveness of the information itself for the simple fact that everyone can see it and, if it is incorrect, directly modify it or ask for its modification. It will be then up to the site operator to check the veracity of the information in case of opposing news.

---

[37] Source: Greenreport, *L'Onu vuole rivoluzionare l'ecoturismo nelle aree protette*, 23 ottobre 2010, Aree protette e biodiversità,
http://www.greenreport.it/_new/index.php?page=default&id=7270
(23 ottobre 2010).

Introducing this "open" system for the collection and verification of information, also scientific and geographical databases store data in large quantities with extreme ease and speed.[38]

Today, the presence of several websites concerning geographic and cartographic information on protected areas is creating some problems with the certainty of the data. The MEDPAN, whose history and organizational structure will be analysed later, is trying to unify the major informational sites by the MAPAMED project.

In 2010, the picture that "Protectedplanet.net" gave of the Adriatic was still incomplete but it had an important base to work on. On that base, thus, a lot of information was stored and many of the starting results verified. The information was then integrated consulting other databases and verifying the results on bibliographical materials available also in printed format in traditional libraries.[39]

---

[38] The project "Ornitho" is well-known in the context of protected areas. It was promoted by a coordination of scientific institutions and it made possible to enter information online on the sighting of bird species according to the place of sighting. By its own website, www.ornitho.it, the research team has extended in an enormous and exponentially increasing way the number of information they had collected thanks to the *birdwatching* lovers of all over the world.

[39] These are the main libraries visited for a deep consultation:
- Central Library - University of Teramo, TERAMO-ITA.
- Library of the Department of Economics and History of the Territory –University "Gabriele D'Annunzio" of Chieti-Pescara. Viale Pindaro, PESCARA-ITA.
- Library of Federparchi- Italian Federation of Parks and Reserves. Via Cristoforo Colombo, ROMA-ITA.
- Library of the Military Geographical Institute. Via Cesare Battisti, FIRENZE-ITA.
- Library of the Italian Geographical Society, Villa Celimontana, ROMA-ITA.
- Center of Environmental Documentation of the National Park of Gran Sasso and Monti della Laga, MONTORIO VOMANO (Te)-ITA.
- Municipal libraries of Roseto degli Abruzzi and Pineto (Te)-ITA.

By examining this documentation, it was possible to select, from a set of more than 200 protected areas recorded on official databases, a large group of protected areas on which to conduct further studies.

Tab.2.A

Coastal and marine protected areas in the Adriatic region

Nazione	Protected Areas
International waters	1
Albania	9
Bosnia Herzegovina	1
Croatia	18
Greece	5
Italy	41
Montenegro	8
Slovenia	4
Total	**87**

87 protected areas have been surveyed in the Adriatic sea and coast by verifying the existence of their regulation establishing measures. The detailed list of protected areas identified with their year of establishment is shown in Tab.2.B.

This is the complete list of the sources:
- **IUCN 2005**: IUCN, *Marine Protected Areas in the West Mediterranean*, SUI July 2005.
- **IUCN 2008**: IUCN-MedPAN-WWF, *Status of Marine protected Areas in the Mediterranean Sea*, SUI 2010.

- Library of CRESA – Abruzzo Centre of Economic Research and Studies, L'AQUILA-ITA.
- Library of the Study Centre of the Museum of Abruzzo Peoples, PESCARA-ITA.
- Library of Primorska University of Koper (Capodistria) SLOVENIA.
- Library of the Marine Protected Area of Miramare, TRIESTE-ITA.
- Library of the Conero Regional Park. SIROLO (An)-ITA.

- **MedPAN 2010**: MedPAN Database on Mediterranean protected Areas, (www.medpan.org).
- **Parks.it 2010**: Federparchi, *L'Italia dei Parchi*, ITA 2010 (www.parks.it).
- **UNDP.GEF 2005**: UNDP/GEF "COAST" Project, *Conservation and Sustainable use of Biodiversity in the Dalmatian Coast*, Center for Coastal Resources Management – Virginia Institute of Marine Sciences, USA February 2005.
- **UNEP 2007**: UNEP, *Report of the II phase of the project "Establishing Emerald network in Montenegro"*, November 2007.
- **BirdLife International 2010**: *IBA Data Base* BirdLife International, (www.birdlife.org).
- **WDPA 2009**, *World Database Protected Areas*, IUCN-WCPA (www.wdpa.org).
- **IZS A&M 2006,** *Guidelines and Management of the Areas of Biological Protection*, Abruzzo&Molise Zoo-prophylactic Institute, 2006.
- **IRSNC:** Institute of the Republic of Slovenia for Nature Conservation - Slovenia
- **MATTM**: Ministry of the Environment and Protection of Land and Sea - Italy
- **MIPAF**: Ministry of Agricolture, Food and Forestry Policies- Italy
- **DZZP**: Croatian State Institute for Nature Protection – Croatia
- **BKCG**: Institute of Marine Biology of Kotor – Montenegro
- **ZRS**: Institute of the Republic of Slovenia for the Conservation of Nature – Slovenia.

2.2. Direct surveys based on interviews

A work to refine the list of 87 marine and coastal Adriatic protected areas started and led to a further categorization to assess whether the activity of the protected area could potentially mean something in terms of opportunities.

This work was carried out through punctual interviews. Questionnaires were sent to a large group of stakeholders and the collected results made possible to get a picture of 41 protected areas of which we were able to know much more detailed information.

Tab.2.B Year of establishment of the Adriatic Protected Areas

#	Area	Year	#	Area	Year
1	Area Pomo ZTB (Deep Sea)	1998	43	Area open Sea (Ravenna)	2004
	ALBANIA		44	Paguro wreck (Ravenna)	1995
2	Karaburun/Vlore	1968	45	Area Barbare (Ancona)	2004
3	Kune-Vain	1960	46	Torre del Cerrano	2009
4	Patok-Fushe-Kuqe	1962	47	Torre Guaceto	1991
5	Butrinti	2005	48	Porto Cesareo	1997
6	Pisha e Divjakes	1966	49	Delta Po Veneto	1997
7	Pishe Poro/Fier	1958	50	Delta Po Emilia Romagna	1988
8	Rushkull	1955	51	San Bartolo	1994
9	Velipoja	1958	52	Conero	1987
10	Narta Lagoon	2000	53	Foce Isonzo	1996
	BOSNIA Herzegovina		54	Foci Stella	1996
11	Mediteranium u Neumu	1965	55	Sentina	2004
	CROATIA		56	Borsacchio	2005
12	Brijuni	1983	57	Calanchi di Atri	1995
13	Limski zaljev	1979	58	Pineta S.Filomena	1977
14	Malostonski Zaljev	1983	59	Grotta Farfalle	2007
15	Mljet	1960	60	Ripari Giobbe	2007
16	Telascica	1988	61	Punta Acquabella	2007
17	Kornati	1980	62	S.Giovanni in Venere	2007
18	Krka	1985	63	Lecceta Sangro	2001
19	Biokovo	1981	64	Punta Aderci	1998
20	Vransko Jezero	1999	65	Marina di Vasto	2007
21	Lastovo	2006	66	Lama Balice	1992
22	Cres-Losinj	2006	67	Duna Torre Canne	2002
23	Velebit	1981	68	Salina Punta Contessa	2002
24	Paklenica	1949	69	Palude Bosco Rauccio	2002
25	Lokrum	1948	70	Costa Otranto S.M.Leuca	2006
26	Prevlaka	2000	71	Litorale Ugento	2007
27	Dio Otoka Krka M. Luka	1969	72	Litorale P.ta Pizzo	2006
28	Dvlije Masline O. Pagu	1963	73	Porto Selvaggio	2006
29	Rt Kamenjak	1996	74	Le Cesine	1980
	GREECE		75	Saline Margherita Savoia	1977
30	Zakynthos	1999		MONTENEGRO	
31	Ammoudia-Loutsa	2001	76	Kotorsko-Risanski bay	1979
32	Etniko Parko limnothalasson	2006	77	Island Katici &Donkovom	2010
33	Vatatsa-Divari-Ormos Valtou	2003	78	Plaza Pecin	2010
34	Kotychi lagoons	1975	79	Buljarica	2010
	ITALY		80	Veliom Hridi &Old Ulcini	2010
35	Gargano	1991	81	Platamuni	2010
36	Isole Tremiti	1989	82	Velika Ulcinjska plaža	2010
37	Area Tremiti	2004	83	Bojana River Delta	1973
38	Miramare	1986		SLOVENIA	
39	Area Miramare	2004	84	Cape Madona	1990
40	Area Tegnùe P.to Falconera	2005	85	Debeli Rtič	1991
41	Area Tegnùe Chioggia	2004	86	Strunjan	1990
42	Area open Sea (Chioggia)	2002	87	Secoveljske soline	2001

In the light of the difficulties of physical or cultural communication between one Adriatic side and the other, in terms of language but also on the interpretation of the concept of protected area, this survey work was carried out entirely via *e-mail*, and only rarely by some personal contact or telephone. It was prepared a closed questionnaire in order to facilitate the users to answer the questions in the form of information sheet for each protected area already surveyed.

The questions were about information on the recognition of the site as a protected area, its year of establishment, its category in the international classification, the type of protected area in the national legislation, the management authority, the number of employees and if there was a plan just for the protection or also for the management and economic planning of the protected area.

The results came mainly from abroad and the information provided was not sufficient to conduct a survey of effective management skills because many did not answer the question on the existence of a management plan or on the number of employees.

So an interesting general picture went out, which is summarized in Tab.2.C, on a number of 41 Adriatic coastal and marine protected areas that already has a good detail of information to be able to start an argument on the management activities.

But before looking at the data of this research to analyze the ascertained situation, it is necessary to study the concept of Network as it t used in the world of nature protection and in the international context of protected areas. Such a study is particularly important for the Adriatic region, and is also an important support to a further analysis in the research and to store even more data on this sample of protected coastal and marine areas which is the Adriatic Network.

Tab.2.C Adriatic Protected Areas with Management Authorities

Id	Site Name	National Designation	Management Authority	Marine or terrestrial	Total Area (ha)	Born	Plan
1	Area Pomo	Fishing Reserve	MIPAF-Ita	Marine	220.000,00	1998	no
2	Narta Lagoon	International Bird Area	Adriatic Center	Coastal	4.180,00	1999	no
3	Brijuni	National Park	Park Authority	Both	3.385,00	1983	yes
4	Mljet	National Park	Park Authority	Both	5.375,00	1960	yes
5	Telascica	Nature Park	Park Authority	Both	6.706,00	1988	yes
6	Kornati	National Park	Park Authority	Both	21.633,00	1980	yes
7	Krka	National Park	Park Authority	Coastal	11.100,00	1985	yes
8	Biokovo	Nature Park	Park Authority	Coastal	19.550,00	1981	no
9	Vransko Jezero	Nature Park	Park Authority	Coastal	5.700,00	1999	no
10	Lastovo	Nature Park	Park Authority	Marine	19.583,00	2006	no
11	Velebit	Nature Park	Park Authority	Coastal	200.000,00	1981	yes
12	Zakynthou	Nat. Marine Park	Park Authority	Both	10.340,00	1990	yes
13	Etniko Parko (lepanto)	National Park	Park Authority	Coastal	33.470,00	2006	no
14	Periochi erivallontikou	National Park	Park Authority	Coastal	152.269,60	2008	no
15	Gargano	National Park	Gargano Park	Coastal	118.144,00	1991	no
16	Isole Tremiti	Marine Reserve	Gargano Park	Marine	14.66.00	1989	no
17	Miramare	Marine Prot.Area	MATTM-WWF	Both	320	1986	yes
18	P.to Falconera (Càorle)	Fishing Reserve	Municipality	Marine	9.000,00	2005	no
19	Area Tegnùe Chioggia	Fishing Reserve	Municipality	Marine	16.022,49	2004	no
20	Torre del Cerrano	Marine Prot.Area	Consortium	Marine	3.700,00	2009	no
21	Torre Guaceto	Marine Prot.Area	Consortium	Both	3.327,00	1991	yes
22	Porto Cesareo	Marine Prot. Area	Consortium	Marine	16.654,00	1997	yes
23	Delta Po Veneto	Regional Park	Park Authority	River delta	12.592,00	1997	yes
24	Delta Po Em.Romagna	Regional Park	Park Authority	River delta	53.653,00	1988	yes
25	San Bartolo	Regional Park	Park Authority	Coastal	1.596,33	1994	yes
26	Conero	Regional Park	Park Authority	Coastal	6.011,00	1987	yes
27	Sentina	Regional Reserve	Municipality	Coastal	177,75	2004	yes
28	Calanchi di Atri	Regional Reserve	Municipality	River	390	1995	yes
29	Pineta S.Filomena	National Reserve	MIPAF	Coastal	19,72	1977	no
30	S.Giovanni in Venere	Regional Reserve	Municipality	Coastal	150	2007	no
31	Lecceta Sangro	Regional Reserve	Municipality	Coastal	165	2001	yes
32	Punta Aderci	Regional Reserve	Municipality Vasto	Coastal	285	1998	yes
33	Lama Balice	Regional Park (Puglia)	Provincia Bari	Coastal	502	1992	no
34	Le Cesine	Regional Reserve	Municipality	Coastal	384	1980	yes
35	Saline Margh.di Savoia	National Natural Park	MIPAF	Coastal	3871	1977	no
36	Kotorsko-Risanski bay	UNESCO Site	Kotor municipality	Both	2.778,79	1979	yes
37	Island Katici	Marine Protected Area	Management planning for these 3 to be in 1 MPA	Marine	439,75	2010	no
38	Plaza Pecin	Nature Monument		Both	153,41	2010	no
39	Buljarica	Nature Monument		Both	302,01	2010	no
40	Strunjan	Natural Park	Park Authority	Both	428,6	1990	yes
41	Secoveljske soline	Natural Park	Soline d.o.o.	Coastal	650	2001	yes

Source: Direct Survey questionnaire based on data taken from the sources referred to Tab.2.B

3. Working in a network: MEDPAN and AdriaPAN

We have already mentioned in the first part of this work the importance of protecting areas of the planet in sufficient numbers in order to represent all primary habitats of life. We have already marked how the dynamic and continuous transformation of ecosystems make everything enormously more complicated.

Thus, we must necessarily resort to specifically designed forms of protection.

In the world the main protection policies in this regard point decisively towards the establishment of protected areas of large surface area and, alternatively or even simultaneously, to the activation of forms of *Networks* of protected areas based on eco-systemic approaches.

The experiences of the 60^s and 70^s in Anglo-Saxon countries were the guidance for applying the formulas of integrated management on a large scale to the protection of marine and coastal environments. The process of establishment of the Great Barrier Reef Marine Park[40] in Australia and, even more important, the activation in the United States of the National Marine Sanctuaries Programme,[41] show such a richness and variety to be unrivaled for both the modernity of the mechanisms by which the multiplicity of territories involved has encouraged the participation of all citizens, and because the policy of coastal and marine environment integrated protection is, in those countries, the ordinary way of managing

---

[40] The *Great Barrier Reef Marine Park*, which can be defined the most famous Marine Park in the world and 345mila Kmq wide, was established in 1975 by a specific law: the *Great Barrier Reef Marine Act*.

[41] The *National Marine Sanctuary Programme* was approved in 1972 by a specific regulatory act, the *Marine Protection, Research and Sanctuary Act*, and has led over time to establish a system of 14 Marine Sanctuaries. The smallest sanctuary in

the territory for a primarily cultural, but then also technical and implementation awareness which goes back a long time ago (Zoppi, 1993).

In the Mediterranean, the issue of the establishment of large protected areas was addressed for the first time in the debate on the aforementioned *Barcelona Convention* and led to the development between 1982 and 1999 of a specific Protocol,[42] which identifies also in the Mediterranean international waters the possibility of establishing the best forms of protection on a large scale: the so-called SPAMI: *Special Protected Areas of Mediterranean Importance.*[43]

An example of this is the relatively recent establishment of *"Pelagos"*, the Sanctuary for the protection of Mediterranean marine mammals, which identifies the Ligurian Sea as a single large offshore SPAMI[44] after an agreement between France, Italy and the Principality of Monaco on 25 November 1999. At the base of this institution there is the commitment of the three countries to protect marine mammals and their environments, protecting them from direct or indirect negative impacts of

---

the Samoa islands has an area of nearly 1 Kmq and the largest is 350,000 Kmq including the Hawaiian Islands, Source: www.sanctuaries.noaa.gov (10.01.2010).

[42] The Protocol, which in its final version bears the name of *Protocol concerning Specially Protected Areas and Biological Diversity* in the *Mediterranean*, was signed for the first time in 1982, together with the other parts of the Barcelona Convention, but in that particular part it was replaced in 1995 by a text specifically concerning biodiversity. That text was signed again in Barcelona. The Protocol entered into force in 1999 after ratification by the appropriate number of individual states.

[43] Cfr. UNEP (1995), *Protocol concerning Specially Protected Areas and Biological Diversity in the Mediterranean (Barcelona-1995)*, Regional Activity Center for Special Protected Areas- CEDEX, Tunis (Tunisia).

[44] The agreement between the three countries was signed in Rome on 25 November 1999, ratified by Italy the following year, on 11 October 2001, and published in the Official Gazette 253 of 30 October 2001.

human activities, in an area of nearly 90,000 square kilometers including Corsica, the Tuscan archipelago and Maddalena, the Ligurian marine protected areas and the Parc de Port Cros. Unfortunately even today the start of the active management of the sanctuary remains problematic since the management of protected areas on the high seas is rather complex and diverse because it includes also the control of maritime traffic and the direct monitoring of international waters (Relini, 2007).

In this regard, it has been observed that, if these issues of control and surveillance at sea were not so complex, much of the Mediterranean Sea, covering an area of over a million and a half square kilometers, it could already be considered one of the most extended "Marine Sanctuaries" of the planet, because the bottom trawling has been banned beyond the depth of 1,000 meters since 2005 by the *General Fisheries Commission for the Mediterranean* (FAO Cattaneo Vietti and Tunesi, 2007).

3.1. Networks among protected areas

If we can intervene on the high seas with unitary and generalized measures, it is harder to do it within the territorial waters or along the coast, where the diversification of interests and skills greatly complicates the implementation of distinct safeguard measures. In these cases, the action that is proving to be more effective for the protection of a wide variety of *habitats* and species is the building of ecological networks of protected areas. This is identified as an integrated, ecological and managing system among different realities sharing the same objective: the *Networks*.

The so-called *Networks* are one of the most interesting experiences that the world of scientific research, and not only, is experiencing in recent years. They are stable networks built among experts and/or institutions sharing the same goals or the same policy areas. It is the new frontier, the most advanced system of coordination of research activities as well as sharing of management experiences.

In 2002, it was mentioned for the first time in official documents at *the World Summit* on Sustainable Development in Johannesburg, while at the "*Marine Summit*" in Washington in 2007, the IUCN (IUCN, 2008) referred to the networks of marine protected areas as a useful system to improve the resilience of marine ecosystems.

> «*The networks can contribute to sustainable development, promoting the integrated marine and coastal management through three related uses and benefits:*
> *A-Ecological: a network can help to maintain the operating of marine ecosystems, including the temporal and spatial scales of ecological systems.*
> *B-Social: A network can help to resolve and manage conflicts on the use of natural resources.*
> *C-Economic: a network can facilitate the efficient use of resources*».

In the last census, carried out by the UN body for monitoring activities on biodiversity conservation, 65 networks of marine protected areas were identified worldwide, including 30 national and 35 international networks, 20 extended on the size of the geographical region and two, including those promoted in the Mediterranean, built by the European Community itself: *Natura 2000* and *Emerald* (UNEP-WCMC, 2008).

The latter two instruments designed within the EU, the first, *Natura 2000*, to be applied directly to the Member States and the second, *Emerald*, intended for non-EU countries, are the most important example at international level of eco-systemic planning which was brought forward to the purpose of creating ecological networks.

At present, of 26,406 sites belonging to the *Natura 2000* network, just 2,612 are related to marine environments.[45] But the number and, therefore, the potential effectiveness of the network, will greatly increase in the next years thanks to the policies implemented through the already examined *EU Strategy* for the *marine environment* and thanks to the future integration of *Emerald* to *Natura 2000*.

In the opinion of some experts, effective coordination is still lacking on this front. That organic connection that should exist among *Natura 2000* sites and the provisions of the national legislation on Protected Areas is still lacking. It is necessary an overall view, able to warrant a more organic and effective protection of marine biodiversity at national and basin level, as it is already a reality in countries such as Germany or the United Kingdom (Tunesi, 2010).

Networks of protected areas that arise spontaneously are even more interesting than Institutional Networks, because of the participation of all stakeholders and the sharing of objectives and strategies underlying their formation.

These forms of self-organized networks are characterized by exchange of resources and means, aimed at solving problems and creating opportunities, without being anchored to the sovereignty of a single state but oriented to build a system of *"governance"* based on the three principles established by the European Union for an actual achievement of a "Sustainable Development": *partnership*, participation and subsidiarity (Gemmiti, 2009).

Someone also mentioned a *"new governance"* as a second stage of the concept of power-sharing government in relation to such a more spontaneous approach as in those protected areas, based on different instruments from legislative ones and more aimed at coordination, learning and participation.

---

[45] The figure of 2,627 is the sum of the marine SCI and SPAs established until 2012. See European Commission-DG Environment, barometer June, 2012, in Natura 2000 n.33 January 2013 pp.8-9.

In a coastal and marine context, the most important network of protected areas, which also concerns Italy, is certainly the one identified by the acronym *MedPAN - Mediterranean Protected Areas Network*. Created in 1990 and re-launched in the early 2000^s on initiative of WWF France, *MedPAN* has come to count over 50 members and 30 partners managing coastal and marine protected areas in 18 countries of the Mediterranean.[46]

3.2. The MedPAN Network

The MedPAN, the network of managers of the MPAs in the Mediterranean was created in 1990 with support from the World Bank (Abdulla, 2008). Its two main objectives were the exchange of experiences among the managers of MPAs and the development and improvement of management tools.

The MedPAN network operated from 1990 to 1996 with seminars and technical and scientific publications. The lack of funds and human resources made the *network* ineffective since 1996, but its value was reaffirmed by the United Nations in 1999 by the RAC/SPA. The National Park of *Port Cros* required a new Statute for MedPAN in 1999, turning it into a non-profit association under the French law, with administrative offices housed in the facilities of *Port Cros*. The RAC/SPA provided the secretarial services for the association, and the responsibility of the network was taken on by the National Park of *Port Cros* and by the French Federation of Regional Parks. The charter of the new association clearly stated its vocation to the Mediterranean network of MPAs:

---

[46] The network of marine protected areas in the Mediterranean was born in 1990 thanks to the efforts of the World Bank, but for many years its actions were limited. In 1999, the *National Park of Port Cros*, which also includes the marine areas of the renowned French Mediterranean archipelago, promoted its transformation along with WWF France for activities that had developeded in the European programmes *Interreg*. Data: www.medpan.org (28.05.2013).

- to increase the flow of contacts and experience among managers of marine and coastal protected areas;
- to attend the training of managers;
- to make available the know-how acquired by some operators to other operators, with a vision of sustainable development;
- to develop and support concrete actions for the planning, management and public awareness of protected areas and protected area networks;
- to improve the development of marine and coastal protected areas, according to the experience of each protected area (Plant, 2003).

In 2001 the National Park of *Port Cros* proposed to WWF France to take on the MedPAN coordination and fundraising. A feasibility study conducted in 2003 refocused attention on the MPAs. From 2005 to 2007, WWF France developed and coordinated a three-year project, funded by the EU through the *Interreg IIIC South Initiative*. It brought together 23 partners from 11 countries of the Mediterranean coasts of which 14 partners are European and 9 non-European. These partners are managing more than 20 MPAs, have worked for the establishment of a wider network tending to include other marine protected areas. During the project, every year several seminars on technical-management subjects which interest all participants were held and management studies and methodological tools elaborated.

After many projects developed thanks to the award of European funding, in 2008, at the IUCN *World Conservation Conference* in Barcelona (Spain), the MedPAN was formally constituted as a non-profit organization legally recognized at the international level, with a new charter and with the purpose of becoming a stable organization that operates through forms of self-financing.

3.3. The AdriaPAN Network

An additional *Network* built many years later had the same shared path and only recently is developing in the Adriatic region, as reported in its certificate of incorporation, the "Charter of Cerrano". It is identified with the name AdriaPAN - *Adriatic Protected Areas Network*.[47]

In the spring of 2008, in Pineto (Teramo) in the establishing "*Torre del Cerrano*" Marine Protected Area, during some seminars addressed to the operators of Marine Protected Areas, the will to be constantly coordinated in territorial areas emerged. From that moment of encounter, the marine and coastal protected areas of the Adriatic Sea started to build a joint network working under the coordination of the *Miramare* Marine Protected Area and the Management Consortium of the *Torre del Cerrano* Marine Protected Area.

On July 8, 2008, another meeting, again in Pineto (Teramo) was attended by representatives of almost all the protected areas of the Italian coast, national environmental organizations, major research institutes and the most involved universities. They met in the form of self-convocation, coordinated by AIDAP *(Association of Italian directors and officers of Protected Areas)* to draw up a document on shared commitments of coastal and marine protected areas in the Adriatic.

A need, felt for years, to participate in joint form to national and international research and to the exchange of useful information on forms of management.

---

[47] AdriaPAN was born on 8 July 2008 in Pineto (Teramo), at the headquarters of the establishing Torre del Cerrano Marine Protected Area and was ratified by the signing of the "Charter of Cerrano" on September 26, 2008 in Rosolina (Ro), in the Veneto Park of the Po Delta. In 2013 the coordination is made up of 40 Coastal and Marine Protected Areas and 43 Adriatic organizations such as universities, research institutes, organizations and governments, associations and committees, etc. Source: http://triviadicerrano.blogspot.com (10.06.2013).

The themes and issues addressed in Pineto concerned the needs of nature conservation and the sharing of administrative experience related to sustainable development in fisheries and tourism.

The meetings ended up with the drafting of a document of intent that was called "Charter of Cerrano" in honour of the place where it was conceived but also to highlight how this document represents the will of everybody that works daily in the field and feel the need to coordinate their work with new realities and different cultures. The aspect, in fact, that most affected the work of Pineto was the interest of several parties, and even of governments of different political colour in the new reality of the Marine Protected Area still in the process of establishment and how this was considered as an important element to appear on the international scene.

Two months later, in the conference room of the *Coastal Botanical Garden* of Porto Calero in Rosolina (Rovigo), in the Veneto Regional Park of the Po Delta, on September 26, 2008, the final version of the *Charter of Cerrano* was ratified on the basis of a draft, drawn up after an intensive coordination.

The final text, fully reported in the Annex, is a set of shared values, objectives and strategies to be pursued, in order to implement a direct collaboration among all marine and coastal protected areas, of any type and form, exclusively in the Adriatic Sea.

In the first ratification the *Charter of Cerrano* was signed by at least one protected area for each Italian region that overlooks the Adriatic Sea. Since then, the subscription is open to all organizations that have to deal with the management of areas in which there is a form of protection oriented to the conservation of biodiversity according to indications coming from the international context. These are the protected areas that signed from the first moment the *Charter of Cerrano* and are therefore the 10 founders of AdriaPAN:

1) *Miramare* Marine Protected Area, Trieste;
2) *Torre del Cerrano* Marine Protected Area, Pineto-Te;
3) *Delta del Po* Veneto Region Nature Park, Rovigo;
4) *Delta del Po* Emilia Romagna Region Nature Park, Ferrara;
5) *Tegnue di Chioggia* Biological Protection Zone, Venezia;
6) *Sentina* Marche Region Nature Reserve, Ascoli Piceno;
7) *Calanchi di Atri* Abruzzo Region Nature Reserve, Teramo;
8) *Lecceta di Torino di Sangro* Abruzzo Region Nature Reserve, Chieti;
9) *Grotta delle Farfalle* Abruzzo Region Nature Reserve, Chieti;
10) *Torre Guaceto* Marine Protected Area, Brindisi.

Two bodies, performing the role of technical support to the organization, also joined it: the AIDAP - *Italian Association Directors and Officers of Protected Areas* and WWF MedPO - *Mediterranean Programme Office*.

Along the whole Italian coast, together with *Slovenia*, *Croatia*, *Bosnia and Herzegovina*, *Montenegro*, *Albania*, as far as *Greece*, all the protected areas freely joined the network by signing the *Charter of Cerrano* and sharing its values and objectives.

As the time passed, AdriaPAN - *Adriatic Protected Areas Network* began to take shape, a stable network among the Adriatic marine and coastal protected areas agreeing with the principles and objectives stated in the *Charter of Cerrano*.

To 2013 AdriaPAN has 40 protected areas, of all the nations that border the Adriatic and signed the *Charter of Cerrano*, and 43 other organizations (Committees, Associations, NGOs, agencies, research institutes, universities, governments, etc..) which joined AdriaPAN sharing its principles and objectives and joining partnerships oriented to the activation of joint development programmes.

In fact, in the light of the great interest that the *network* has aroused in research institutes, universities, associations and other stakeholders, the same protected areas that are part of AdriaPAN have allowed access to the network to other entities which adhere to AdriaPAN sharing also the principles and objectives of the *Charter of Cerrano*.

In 2013, this is the list of **signatories** of the *Charter of Cerrano* and members of the AdriaPAN network:

1) *Torre del Cerrano* Marine Protected Area, Pineto-Te
2) *Miramare* Marine Protected Area (Trieste-Ita)
3) Delta del Po Veneto Region Park (Venezia-Ita)
4) Delta del Po Emilia Romagna Region Park (Ravenna-Ita)
5) Torre Guaceto Marine Protected Area (Brindisi-Ita)
6) Sentina Region Nature Reserve (S. Benedetto Tronto-AP-Ita)
7) Calanchi di Atri Region Nature Reserve (Atri-TE-Ita)
8) Tegnùe di Chioggia Biological Protection Zone (Venezia-Ita)
9) Lecceta Torino di Sangro Regional Nature Reserve (Chieti-Ita)
10) Grotta delle Farfalle Regional Nature Reserve (Chieti-Ita)
11) Mljet National Park (Dubrovnik-Hrvatska)
12) Kornati National Park (Zadar-Hrvatska)
13) Brijuni National Park (Pula-Hrvatska)
14) Wetland Lagoon of Narta (Vlore-Albania)
15) Marina di Caorle–Tegnùe di P.to Falconera Oasis (Venezia-Ita)
16) Lastovsko otočje Nature Park (Ubli-Hrvatska)
17) Monte San Bartolo Regional Park (Pesaro-Ita)
18) Conero Regional Park (Ancona-Ita)
19) Le Cesine State Nature Reserve (Lecce-Ita)
20) Gargano National Park (Foggia-Ita)
21) Isole Tremiti Marine Protected Area (Foggia-Ita)
22) S.Giovanni in Venere Nature Reserve (Chieti-Ita)
23) Punta Aderci Nature Reserve (Chieti-Ita)
24) Prvic Special Reserve (Rijeka-Hrvatska)
25) Cres Island Special Reserve (Rijeka-Hrvatska)
26) Lopar Important Landscape (Rijeka-Hrvatska)
27) Kolansko Rogoza Special Reserve (Zadar-Hrvatska)
28) Veliko i Malo Special Reserve (Zadar-Hrvatska)
29) Dugi otok Island Important Landscape (Zadar-Hrvatska)
30) Zut-Sit Archipelago Significant Landscape (Sibenik-Hrvatska)
31) River Krka lower course Significant Landscape (Sibenik-Hrvatska)
32) Sibenik Channell-Harbour Significant Landscape (Sibenik-Hrvatska)

33) Neretva River Delta Special Reserve (Dubrovnik-Hrvatska)
34) Mali Ston and Malo More Special Marine Reserve (Dubrovnik-Hrvatska)
35) Island Mrkan, Bobana and Supetar Special Reserve (Dubrovnik-Hrvatska)
36) Saplunara Island Significant Landscape (Dubrovnik-Hrvatska)
37) Badija Island Significant Landscape (Dubrovnik-Hrvatska)
38) Telasčica Nature Park (Zadar-Hrvatska)
39) Strunjan National Park (portoroz-Slovenia)
40) Ripabianca Regional Nature Reserve (Jesi-AN-Ita)

This is the list of the other organizations **adhering** to AdriaPAN:

1) AIDAP Ass. It. Direttori e funzionari Aree Protette- Feltre (Bl) ITA.
2) WWF Mediterranean Programme Office- Roma ITA.
3) SUNCE Ass. for nature, environment and sustainable development-Split-HVR.
4) AULEDA Local Economic Development Agency- Vlore ALB.
5) Adriatik Center - Vlore ALB.
6) Università di Teramo, Dip. Teorie Politiche Sviluppo Sociale-Teramo ITA.
7) Università di Teramo, Dip. di Scienze Biomediche Comparate- Teramo ITA.
8) Università di Bari, Dipartimento di Zoologia- Bari ITA.
9) Un. di Bologna-CIRSA, Centro Interdip. Ricerca Scienze Amb.-Ravenna ITA.
10) Università di Roma 3, Dipartimento di Biologia Ambientale- Roma ITA.
11) IZS A&M "Caporale", Ist. Zooprofilattico Abruzzo&Molise -Teramo ITA.
12) Consorzio Mario Negri Sud, Environment Rsearch Center- Lanciano (Ch) ITA.
13) Time Project, EU Project development Bolzano.
14) Centro Studi Cetacei, Association - Pescara ITA.
15) Tethys Research Institute- Milano ITA.
16) Blue World Institute of Marine Research and Conservation- Split HVR.
17) Morigenos, Marine mammal research and conservation society- Piran SLO.
18) Fond. Cetacea Onlus -Riccione ITA. -Reef Check Italia Onlus -Ravenna ITA.
19) Consorzio CIVICA, EU Project development-Pescara ITA.
20) ITACA, Association for Local Development-Teramo ITA.
21) Coordinamento Tutela Costa Teatina - Chieti ITA.
22) Natura Jadera, Pub.Inst.for Manag.of Nature P.A. in Zadar County- Zadar HVR.
23) Sibenik Nature, Pub.Inst. for Management of P.A. in Knin County- Sibenik HVR.
24) Priroda, Pub.Inst. Priroda County of Primorje and Gorski Kotar- Rijeka HVR
25) Dubrovnik-Neretva Nature - Dubrovnik HVR
26) SELC, Società per l'Ecologia delle Lagune e delle Coste- Venezia ITA.
27) EcoVie - Chieti ITA.
28) Comitato Riserva Nat. Reg. Guidata Borsacchio- Roseto degli Abruzzi (Te) ITA.

29) Diatomea, Environment planning- Senigallia (An) ITA.
30) Archeosub Hatria, Association- Silvi (Te) ITA.
31) LANDS, Professional network- Maranello (Mo) ITA.
32) Monk Seal Group, Association- Roma ITA & Pula HVR.
33) BluMarine Service, scarl- San Benedetto Tr. (Ap) ITA.
34) Nature Survey, Association- Milano ITA.
35) HabitatLAB, Onlus- Pescara ITA.
36) Nautilus, NGO- Kotor CRG.
37) Green Home, NGO- Podgorica CRG.
38) Dolphin Biology&ampConservation, Scientific staff- Perugia ITA.
39) Ocean Care NGO - Wädenswil, SUI.
40) Camera Commercio Tirana - Tirana ALB.
41) EURAC research - Bolzano ITA.
42) MedCEM, Mediterranean Center for Environment Monitoring, NGO - Bar CRG.
43) CRASsrl, Centro Ricerche Applicate Sviluppo Sostenibile- Roma ITA.

AdriaPAN starts as a *bottom-up* initiative creating those conditions of stable cooperation among the operators of protected areas in bordering countries which, unfortunately, is not easy to achieve through their ordinary diplomatic channels. AdriaPAN moves in the wake of the positive experience of MedPAN that between 2006 and 2009 was the main *network* of marine protected areas in the Mediterranean. The success recognized in that network system, although it is still young, is linked to its basic formula involving the individual managers of protected areas and promoting their real interest in working together on common goals. On October 6, 2008 in Barcelona, Spain, the IUCN held the *IV World Conservation Conference - International Union for Conservation of Nature.* In this context, as part of the MedPAN, AdriaPAN was presented on the international scene. The occasion was the meeting organized by the *Nature Conservancy*, Federparchi (Italian Federation of Parks and Nature Reserves) and WWF France, entitled "*Speeding up the establishment of a coherent, representative and effectively managed ecological network of marine protected areas in the Mediterranean?*".

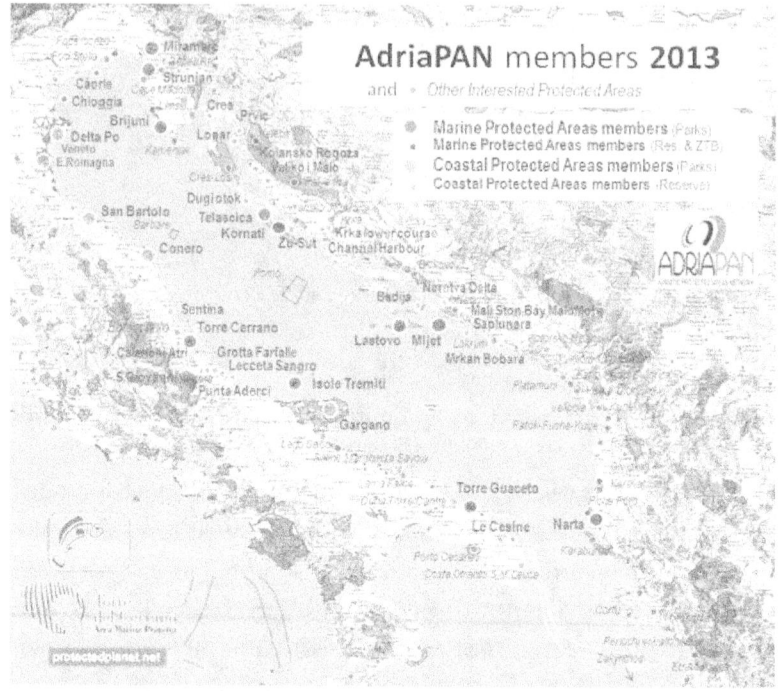

*Fig.3.a The Marine and Coastal Protected Areas in AdriaPAN Network
(Source: http:/triviadicerrano.blogspot.com)*

Since then AdriaPAN has officially started its activities in the international context and on the basis of its charter, the *Charter of Cerrano*, has enabled the preparation of projects of common interest for all coastal and marine protected areas of the Adriatic Sea.

Still in 2013 AdriaPAN is only a coordination of managers of Italian, Slovenian, Croatian, Montenegrin and Albanian coastal and marine protected areas in the Adriatic Sea, and works on the action of each unit operating in the various protected areas with a minimum of coordination by the promoters of the initiative: the Miramare and Torre del Cerrano Marine Protected Areas.

To 2013 AdriaPAN can already boast good results despite the few years of activity. There have been several major awards and twelve projects on an international scale are being developed.

The first recognition, at a political-institutional level, was expressed by the AII - *Adriatic & Ionian Initiative*, the above-mentioned organization led by the foreign ministries of the Adriatic countries. During the meeting of the *Senior Officials* held in Ancona on March 26, 2010, AII gave its patronage to the activity of AdriaPAN inserting it among the existing networks and structured organizations to refer to for activities of cooperation. The networks of Adriatic Universities *(UniAdrion)*, Adriatic Cities and the Forum of the Chambers of Commerce have had the same recognition as the AdriaPAN network of protected areas.

The second result, of technical-organizational type, occurred when the AdriaPAN Secretariat was called to be part of the *Steering Committee* established by MedPAN to process the database of the Mediterranean marine protected areas on behalf of the IUCN. Two meetings on this topic were held in September 2010 in Marseille in France and in November 2010 in Korba in Tunisia and AdriaPAN proved with its own experience how considering both Marine Protected Areas and Coastal Protected Areas is important for a realistic conservation action of the Mediterranean ecosystem.

A further and even more recent result was a scientific-academic recognition occurred when AdriaPAN was invited to participate and take a report on the activities being carried out in the workshop organized by UNEP[48] to take stock of the 2012 targets in terms of marine biodiversity.

---

[48] United Nations Environment Programme.

The conference was held in Slovenia on October 29, 2010 at the Institute of Marine Biology of *Piran*, under the title "*Toward a representative network of Marine Protected Areas in the Adriatic*".

The two projects funded to start up an AdriaPAN secretariat (AdriaPAN Secretariat and PANforAMaR) were also important. The first was funded by the call for *Small Project* sponsored by MedPAN with funding FFEM, MAVA, Foundation Albert of Monaco; the second by the call for cooperation projects, published by the AII - *Adriatic & Ionian Initiative*.

Finally, the highest recognition for AdriaPAN was the quote as a good practice to be taken as an example of international cooperation, international cooperation on the marine environment, by the European Commission in the document for the Maritime Strategy for the Adriatic in the European Council Decision of November 30, 2012 (COM 2012-713 final).

3.4. The added value of Networks

Gathering up the marine protected areas of a region, however, does not automatically build a *network*. At the beginning we can speak of an "aggregate" rather than a "conglomerate" [49], an association of administrative organizations, designated often opportunistically, sometimes even with different goals. Neither the geographical proximity of protected areas or collecting them in a single institutional container, although this is recognized in legal terms, are sufficient criteria to give birth to an ecological *network*. In order to make an ecological sense, Networks of protected areas, in particular marine and coastal areas, should be planned to achieve the same goal. A

---

[49] It is well-known the definition given in 2005 by Notarbartolo Di Sciara which refers to systems of MPAs as a «conglomerations of individual MPAs or networks under a multi-institutional framework, strategically planned and operating in a coordinated manner.»

network should be imagined as a subject with a single management plan and in which individual parties act as focal centres of conservation (Spoto, 2009).

The spontaneous emergence of these networks is an obvious consequence of the desire to independently activate the necessary system policies, invoked by many, but still difficult to implement in Italy. It seems obvious, however, that if any institutional support, not only economical but also in terms of operational facilities, does not follow such so-called *"bottom up"* spontaneous initiatives of coordination, it is likely to diminish the initial enthusiasm and the strong group ability that are rather important assets to be capitalized.

The added value of Networks for marine and coastal protected areas has been perceived since, at the IUCN-WCPA *Marine Summit* in Washington in 2007, official documents started mentioning "corridors" of ecological connection among protected areas.

The planning and management techniques designed and used in the protected areas, especially the application of the voluntary instruments for sustainability, would enable, even with relative ease, to develop strategies and shared action plans even in territories so difficult to manage as a unit such as coastal areas. Where, in fact, very different realities interact with one another, those relating to the sea and those to the mainland, different from environmental, socio-economic, administrative and institutional points of view, the importance of using "coded" systems, such as cooperation networks, increases. Stable *Networks* make possible to draw on the experiences of others and to achieve the desired results with greater ease.

Complex protocols designed and approved at the international level, such as the Integrated Coastal Zone Management (ICZM), seem to be the most far-sighted, though, inevitably, those with the greatest difficulties of application if not activated, for now, through a spontaneous adhesion of all involved parties. It is clear in these situations how strong is the desire to create *Networks* among protected areas also in legislation.

Finally, the high planning quality required by the European Community for any type of funding application is unlikely to be reached in the situation in which today most Italian protected areas, especially coastal and marine areas in the Adriatic, are moving. The world of protected areas is strongly supported by the "third sector", non-profit associations, cooperatives and non-profit organization, working with competence, professionalism and passion. The scarcity of economic resources from the States pushes more and more these management realities to rely on EU funding programmes. So the *Network* is a place for exchange experiences and information useful for scientific research and, in this situation, it has also become a context in which it is possible to unite strategically planning forces in order to achieve common goals.

4. A survey on the Adriatic

In 2007 the IUCN[50] started the initiative *"How is your MPA managed?"* with the intent to support directly the Action Plan 2006-2012 of the WCPA-Marine[51], in particular about the statute commitment «to help Governments and other Parties to plan, develop and establish MPAs, networks of MPAs and the global system of environmental protection». Starting from the observation that many protected areas are not able to play a leadership role in the formulation of their management plan, the method outlined in the manual *"How is your MPA managed?"* is complementary to what previously described in the manual *"How is your MPA doing?"*, providing a practical and interactive procedure to guide the

---

[50] International Union for Conservation of Nature.
[51] World Commission on Protected Areas–Marine.

managers of MPAs, together with their staff, representatives of the local community and stakeholders in the definition and development of the management plan. It is reiterated that the performance indicators are an integral part of any management plan, and that, if taken alone, are devoid of any usefulness. Their measurement should therefore be defined in advance and must be integrated in the work plan of the MPA. The results should be analysed, discussed and shared. They will serve as a starting point for the next management cycle.

A first training session on the use of this method, which was held in Barcelona in September 2008 by the authors of the manual *"How is your MPA managed?"* and addressed to twenty representatives of as many MPAs.

It is one of the most interesting experiences for the evaluation of management effectiveness of protected areas, for now only marine ones, to which to refer in this work also looking at the results of the first application launched in Italy.

In 2005, the Association WWF Italy (as the managing body of the MPA Miramare) and Federparchi (Italian Federation of Parks and Nature Reserves) launched the initiative *"Tools for assessing the effectiveness of management and adaptive management for the system of the Italian marine protected areas"* thanks to funding from the Ministry of the Environment, Land and Sea - Directorate of Nature Protection.

Under the guidance of the MPA Miramare, the other Italian MPAs which took part in the project are:

- MPA Tor Paterno (Rome), the only non-Italian coastal reserve, located in the open sea,
- MPA Torre Guaceto (Brindisi), a marine reserve with a State Nature Reserve on land and a RAMSAR wetland,

- MPA Isole Ciclopi (Catania), due to the characteristic of being made up of islands,
- MPA Penisola del Sinis - island of Mal di Ventre (Oristano), as it has both a part of the coast and an island.

The Italian translation of "*How is your MPA doing?*" was the first step in implementing the project. The second step made possible to adapt the translated manual to the Italian context, reinterpreting the original manual, which in some aspects places more emphasis on the marine protected areas in countries of the developing world. So it was contextualized to the current Italian situation, characterized by increasing human pressure and tourism, as well as a higher level of welfare of the people living near the MPA.

The results obtained by the five MPAs have denoted a full maturity of the Ciclopi, Miramare, Sinis, Tor Paterno and Torre Guaceto MPAs from the point of view of the managing results divided up in means and/or services intended for use, storage, communication and information, management of resources, development and local productions. The fruition, communication and information systems are certainly the most developed taking into account that usually 75% of the possibilities of fruition in the surveyed protected areas have been implemented and are operational (visitor centres, educational workshops, posters, mooring fields, nature trails, promotional material, website, etc.). Most of the MPAs denounce, on the other hand, a low level of monitoring, control and management of tourist flows (e.g. waste on the beaches and in the sea).

Regarding conservation, the five MPAs have a good set of tools for conservation equal to 72.5% of those surveyed, a cartographic GIS processing, biological monitoring programmes in zones A and B, they conduct studies on biocoenosis and have already got the environmental certification EMAS. Finally, the promotion of sustainable local productions is good (66.7%), while the resource management is poor (30%) that is the presence of programmes of alternative energy sources, recycling on the coast and at sea, disposal improvement projects. These activities should certainly be encouraged by targeted projects (MATTM, 2007).

The research regarding the Adriatic protected areas in this work started from the instrument just described, conceived at international level and applied and tested in Italy for the above-mentioned five marine protected areas. The present work can be considered then as a further step of applying research to a larger and far more heterogeneous sample of protected areas.

Only the protected areas in the AdriaPAN network in 2011 were selected for this study. Both their management orientation, since they had signed the *Charter of Cerrano*, and references to the managing body, being able to easily find the information within the same network, were already known.

It was prepared a closed questionnaire in order to facilitate the users in answering the questions in the form of an information sheet for each protected area already surveyed.

It was possible to get a result of some interest, as Tab.4.A shows, also by sending *e-mails*, but then only to those directly involved. No more than half of the respondents responded to the questions specifically addressed to the representatives of the MPAs which subscribed the *Charter of Cerrano*.

The selected topics on which the survey was carried out were just a few of those reported in the work promoted by the Italian Ministry of Environment and implemented by WWF Italy and Federparchi.

Tab.4. A Picture of socio-economic data collected in the coastal and marine protected areas of the AdriaPAN network

Protected Area / Area Protetta	Headquarters building / Sede e Uffici	Visitor Center / Centro Visite	Border Signal / Cartellonistica di confine e boe	Information Points / Punti Informativi	Museum / Museo	Touristic Harbor / Porti turistici	Touristic terrestrial path / Percorsi turistici terrestri	Touristic marine path / Percorsi turistici marini	Touristic fishing / Pescaturismo	Over season tourism / Turismo fuori stagione	Touristic specific Plan / Piani Specifici per il Turismo	Touristic Environmental Impact / Impatti ambientali turistici	Protected Area Logo / Logo dell'Area Protetta	Local typical production / Produzioni tipiche locali	Logo used as product mark / Logo utilizzato come marchio	EMAS / Iscrizione EMAS	Visitors per Year (Sign with # who motivated by protected area) N° VISITATORI ANNUI (Segnare con # se motivati da Area Protetta)
AdriaPAN members																	
Narta Lagoon	Y	?	?	?	?	?	-	?	?	?	?	?	?	?	?	?	?
Brijuni	Y	no	no	Y	Y	Y	Y	no	Y	Y	no	Y	Y	Y	Y	no	# 170.000
Mljet	Y	no	no	Y	no	Y	Y	no	Y	no	no	Y	Y	no	no	no	100.000
Komati	Y	no	Y	Y	no	Y	Y	no	Y	Y	no	Y	Y	Y	no	no	# 80.000
Lastovo	Y	no	no	Y	no	Y	Y	Y	Y	no	no	Y	Y	Y	Y	no	# 28.000
Gargano	Y	Y	Y	Y	?	Y	Y	-	Y	Y	?	Y	Y	Y	?	?	?
Isole Tremiti	Y	?	?	?	?	Y	?	?	Y	Y	?	Y	Y	?	?	?	?
Miramare	Y	Y	Y	Y	Y	Y	Y	Y	Y	Y	Y	Y	Y	Y	Y	Y	?
Area Tegnùe Pto Falconera (Caorle)	Y	Y	Y	no	no	Y	Y	Y	Y	Y	Y	Y	Y	Y	Y	Y	not available
Area Tegnùe Chioggia	Y	?	Y	?	?	?	-	Y	?	Y	?	?	Y	?	?	?	?
Torre del Cerrano	Y	no	no	Y	no	no	no	no	Y	Y	no	Y	Y	Y	no	no	?
Torre Guaceto	Y	Y	Y	Y	Y	no	Y	Y	Y	Y	Y	Y	Y	Y	Y	Y	?
Delta Po Veneto	Y	Y	Y	Y	Y	?	Y	-	?	Y	Y	?	Y	Y	?	?	?
Delta Po Emilia Romagna	Y	Y	Y	Y	Y	?	Y	-	?	Y	Y	?	Y	Y	?	?	?
San Bartolo	Y	?	?	?	?	Y	?	-	?	?	?	?	Y	Y	no	no	?
Conero	Y	Y	Y	Y	Y	Y	Y	-	Y	Y	Y	Y	Y	Y	?	?	?
Sentina	Y	no	Y	Y	no	Y	Y	-	Y	Y	no	Y	Y	Y	no	no	not available
Calanchi di Atri	Y	Y	Y	Y	no	no	Y	-	-	Y	Y	no	Y	Y	Y	no	# 3.000
Grotta Farfalle	no	Y	no	no	no	no	no	-	Y	no	no	no	no	Y	no	no	?
S.Giovanni Venere	Y	Y	no	no	no	no	no	-	no	no	no	no	no	Y	no	no	?
Lecceta Sangro	Y	Y	Y	Y	no	Y	Y	-	no	Y	no	Y	Y	Y	no	no	3.000
Le Cesine	Y	Y	Y	?	?	?	?	-	?	?	?	?	?	Y	?	?	?
TOTALI	21/22	12/18	13/19	14/17	6/16	12/17	14/17	5/9	13/16	16/19	6/16	13/16	18/20	18/19	6/14	3/14	
Other Protected Areas as check / Altre Aree Protette per controllo campione esterno																	
Isole Ciclopi	Y	Y	Y	Y	Y	Y	Y	Y	no	Y	Y	Y	no	Y	Y		
Penisola Sinis	Y	Y	Y	Y	Y	Y	Y	Y	Y	Y	Y	Y	Y	Y	no	no	
Secche Tor Paterno	Y	Y	Y	no	Y	Y	no	-	-	-	Y	Y	-	-	-		

Source: Direct survey by a questionnaire developed on the basis of data of the volume Ministry of the Environment and Protection of Land and Sea, *Evaluation of management effectiveness of Italian MPAs*, EUT, Trieste 2007.

Specifically, the closed-ended questions concerned the following topics simply asking if they existed in the protected area: headquarters and offices, visitor centres, border signs and buoys, information points, museums, marinas, terrestrial tourist routes, marine tourist routes, fishing tourism, off-season tourism, specific plans for tourism, environmental impacts of tourism, logo of the Protected Area, typical local products, logo used as a trademark, EMAS registration, number of visitors per year.

Tab.4.B Survey of socio-economic data

Question	Yes	Answer	%
Headquarters and offices	21	22	95,5
Visitor centres	12	18	66,7
Border signs and buoys	13	19	68,4
Information points	14	17	82,4
Museum	6	16	37,5
Marinas	12	17	70,6
Terrestrial tourist routes	14	17	82,4
Marine tourist route	5	9	55,6
Fishing tourism	13	15	86,7
Off-season tourism	15	19	78,9
Specific plans for tourism	6	15	40,0
Environmental impact of tourism	13	16	81,3
Logo of the Protected Area	18	20	90,0
Typical local products	18	19	94,7
Logo used as a trademark	6	14	42,9
EMAS registration	3	14	21,4

Most of the structures at the base of the question are essential for the operation of the protected area, while others are interventions promoted by the management bodies in order to improve the usability by visitors or to make more attractive the image of the protected area on the tourist market. The last two questions are addressed to identify specific management choices that can start self-financing formulas, the question on the use of the trademark, or management effectiveness through the certification process, the question about EMAS registration.

The survey result can only be partial and is not able to describe a real-world overview on the ability of each protected area to be a development opportunity for its territory. Looking at the histogram shown in Tab.4.C with the percentages calculated with respect to the responses received, we can dare some observations.

Tab.4.C Overview of per cent results on responses received (direct survey)

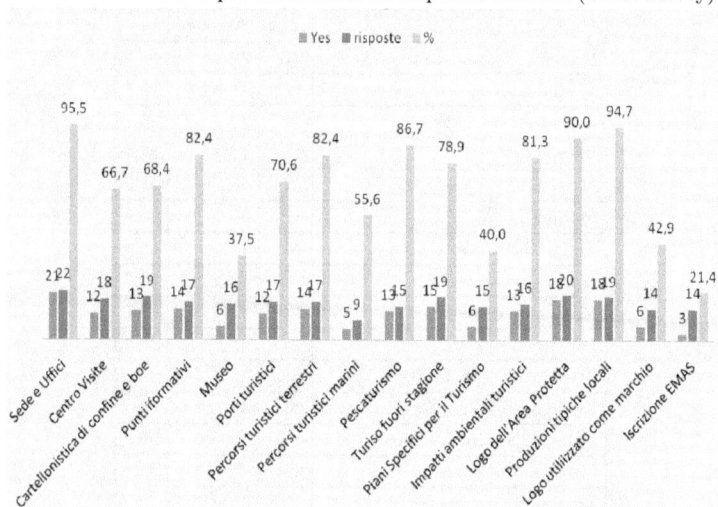

Almost all protected areas have their own headquarter for their organization and management as well as independent operating offices, a logo, that is a brand that identifies the protected area. Only about half of those protected areas, however, has a Visitor Centre along with offices, just over half uses a logo and less than half uses a trademark to identify a local product. Yet all protected areas that responded to the questionnaire have on their territory a public utility for tourism and the production of typical local products.

Few, less than half, are the protected areas that have developed a plan for tourist activity, and there are very few protected areas that have started a form of evaluation and control of their work with standard methods which may be the registration to EMAS.

Conclusions

An extraordinary sea, with shallow bottoms, poor water exchange, rich in biodiversity, home to many Mediterranean life forms and "nursery" for a large number of fish species: this is the Adriatic Sea in its natural aspects. A concentration of cultures that thousands of years of history on its shores has layered leaving presences, finds, monuments, traces and traditions, an irreplaceable heritage of mankind, concurred to this wealth. The Adriatic is a jewel, a precious slice of our planet that should be protected and preserved in its entirety. So much is being done in a geopolitical picture difficult to be managed and a social reality with different languages, cultures and religions. But it is not enough. In the years of "Anthropocene", in particular in the last century, the conditions of the Adriatic have slowly and steadily got worse both in terms of quality, such as the presence of biodiversity and state of conservation of historical and artistic goods, and in quantity; the availability of fish resources have dramatically collapsed and cultural heritage has been lost as the systems of homogenization of national cultures have prevailed on local uses, customs and traditions.

The protected areas in general, parks, reserves, world heritage sites, historic sites, (natural or architectural) monuments, biological protection zones, botanical gardens, restricted archaeological sites, whatever the form of protection, have been the only areas where forms of restriction and/or accurate management systems have helped to maintain and sometimes improve the state of conservation of the existing natural and cultural heritage.

The timely protection, that is recalled by several parties with the metaphor of Noah's Ark, on the side of conservation of natural resources, biodiversity, species in danger of extinction, meets increasing difficulties. Protected areas are increasingly besieged by a use of land, sea and their resources completely incompatible to any principle of sustainable development.

Uncontrolled urban sprawl, hardening of shorelines, lack of water treatment, offshore installations for the extraction or storage of hydrocarbons, forms of fishing which devastate bottom ecosystems, overfishing with the use of new technologies, increase in pleasure boating and tourist activities exceeding the resilience of the environment, the transport infrastructure of goods and/or energy sources with a strong environmental impact, various kinds of abusive activities from building to fishing, from dumping to withdrawals of material.

These are just some of the issues that surround and, unfortunately, sometimes affect the coastal and marine protected areas in the Adriatic. A radical change of direction would be required, a much more controlled growth if not a serene degrowth, but now the complexity of the economic systems of the countries surrounding the Adriatic gives no hope to these expectations. So, now it is important to follow with even greater determination strategies adopted in the international context for the short term and to act in order to achieve the goals set by the CBD and IUCN in the "AICHI Target" for a 10% of protected sea and for a real and effective network of protected areas, efficient and resilient, before 2020.

To maintain a minimum of credibility and dignity towards future generations.

Annex – "Cerrano" Charter

CERRANO CHARTER

The Cerrano Charter sets the basis for the establishment of the Network of Marine and Coastal Protected Areas in the Adriatic Sea, or "AdriaPAN Network": constitution, objectives and interventions.

Text adopted unanimously on July 8th, 2008 in Villa Filiani in Pineto (TE) Italy, "Torre del Cerrano" Marine Protected Area

Definitions

Pursuant to this charter, the terms below have the following meanings:
- **Marine and Coastal Protected Area** means any defined area within or adjacent to the marine environment, together with its overlying waters and associated flora, fauna, and historical and cultural features, which has been reserved by legislation or other effective means, including custom, with the effect that its marine and/or coastal biodiversity enjoys a higher level of protection than its surroundings.
- **Adriatic network** means a group of Marine and/or Coastal Protected Areas maintaining a coherent ensemble of critical habitats necessary for dynamic functioning of ecological processes, which are essential for the biodiversity and regeneration of natural resources in the Mediterranean sea.
- An **Ecoregion** is a large unit of land or water containing a geographically distinct assemblage of species, natural communities, and environmental conditions. The boundaries of an ecoregion encompass an area within which important ecological and evolutionary processes most strongly interact" (WWF 2003). Ecoregion conservation "is an evolution in thinking, planning, and acting at the spatial and temporal scales best suited for successful biodiversity conservation"

Main objective

The main objective of the AdriaPAN network is to initiate a technical process in support of all MPAs managers, aimed at speeding up the achievement of the goal set during the World Summit on Sustainable Development (WSSD)[1] to establish networks of Marine and Coastal Protected Areas by 2012.

---

[1] World Summit on Sustainable Development, Plan of Implementation. 31(c): ("sviluppare e facilitare l'uso di diversi approcci e strumenti, includendo ... la costituzione di aree marine protette

In accordance with such an international commitment, as well as with the Convention on Biological Diversity (CBD), the countries bordering the Adriatic Sea are called upon to halt marine and coastal biodiversity loss through the identification and design of a system of regional networks of ecologically and culturally representative coastal protected areas, managed effectively, and to create the conditions necessary to achieving such a system by 2012.

In compliance with the European Directive 2008/56/CE *("Framework Directive on Strategy for the Marine Environment", 17/6/2008)*, the activation of the Network of Marine and Coastal Protected Areas in the Adriatic Sea - AdriaPAN aligns with:
- The demand for implementation of thematic strategies - which are precisely those undertaken by the Marine and Coastal Protected Areas - for the management of human activities that have an impact on marine and coastal ecosystems;
- The request (Article 13, paragraph 4) to implement conservation measures that can contribute to the establishment of coherent and ecologically representative networks of Marine and Coastal Protected Areas.

Managers of Marine and Coastal Protected Areas and of coastal protected areas whose perimeter is partly in contact with the Adriatic sea,[2] characterized by shared common concerns linked with the protection and proper fruition the Adriatic Sea and coast, voluntarily adhere to the Network of Marine and Coastal Protected Areas of the Adriatic Sea - AdriaPAN. They constitute the "nodes" (nodi) of the network.

The Charter stems from a strong need for coordination among all actions related to the management of Marine and Coastal Protected Areas in the Adriatic Sea.

The Adriatic Sea has always represented an area where different lands and cultures found a common language, new forms of commerce and where, perhaps more than elsewhere, the coastline represented the search for a cross-border cooperation and development. In order to further the socio-economic consolidation and cohesion of the Adriatic ecoregion it is essential to consider the conservation of the marine and coastal environment as the main pillar (structural element) of any plan or programme, with a special emphasis on the most sensitive and vulnerable areas, most notably the

---

in accordo con le leggi internazionali e basate su informazioni scientifiche compresi networks rappresentativi, entro il 2012"...

[2] Secretariat of the Convention on Biological Diversity (2004). TECHNICAL ADVICE ON THE ESTABLISHMENT AND MANAGEMENT OF A NATIONAL SYSTEM OF MARINE AND COASTAL PROTECTED AREAS, SCBD, 40 pages (CBD Technical Series no. 13).

coastline, where most economic activities are concentrated and where natural resources are most compromised.

To ensure the proper management of Marine and Coastal Protected Areas in the Adriatic Sea it is therefore necessary:
- To define priorities for the conservation of key marine and coastal biodiversity features of the Adriatic, including through gap analyses;
- To identify the stakeholders and their level of dependence on the natural resources (such as those operating in the fisheries sector and in tourism);
- To engage local actors, both public and private, in common conservation strategies aimed at promoting sustainable development in and in the vicinity of the protected areas;
- To identify all risks caused by illegal practices and then call on the relevant institutions for their intervention;
- To promote the management of the Marine and Coastal Protected Areas on the basis of common standards and protocols;
- To disseminate knowledge and the comprehension of the environmental, socio-economic and cultural (tangible and intangible) values of the Adriatic Sea and coasts;
- To promote the role of protected areas as an effective tool to achieve national conservation priorities, contribute to the halt of biodiversity loss and the reduction of poverty;
- To disseminate and apply the concepts of integrated coastal and marine areas management;
- To understand the political and administrative processes that lead to the creation of Marine and Coastal Protected Areas and promote the establishment of new Marine and Coastal Protected Areas in order to expand the network, with a special regard to those biodiversity features still underrepresented of environmental protection still little represented;
- To identify funding opportunities in support to the Marine and Coastal Protected Areas and the operation of the network.

The AdriaPAN network is an integral part of the network of Marine and Coastal Protected Areas managers in the Mediterranean – MedPAN. Within such a network, AdriaPAN represents and promotes the ecological, cultural and economic specificities of the Adriatic sea and coasts. The existing networks of protected areas in the Mediterranean, such as MedPAN, the SPAMI list (Special Protected Areas of Mediterranean Importance), the network of Natura 2000 sites, the network of Ramsar sites, the UNESCO Man and Biosphere reserves, and the network of protected areas belonging to Europarc, represent a reference for the AdriaPAN network.

The AdriaPAN network interfaces with existing networks that are representative of the Adriatic that do not address environmental issues exclusively, such as the Adriatic Euroregion network.

The protocols and tools made available by European and international institutions on MPA management and integrated coastal and marine areas management, such as ICZM Recommendation of the European Commission, the European Charter for Sustainable Tourism, the EC Eco-label, EMAS, represent other elements of reference for the future activities of AdriaPAN.

The Marine and Coastal Protected Areas adhering to the Charter of Cerrano commit to actively contributing to the protection of the marine and coastal environment in the Adriatic ecoregion, to safeguarding such habitats and ecosystems and ensuring their good environmental status as well as the flux of benefits to the coastal communities which still depend on natural resources.

Specific objectives

The specific objectives of the network AdriaPAN are, therefore, the following:
- To ensure the networking of Mediterranean MPAs and create synergies between them;
- To promote technical and institutional exchange between members of the network;
- To facilitate the implementation of concrete actions of common interest to managers, able to contribute to the improvement of Mediterranean MPA management, particularly through improving the dissemination of knowledge and management techniques of these protected areas;
- To facilitate fund raising for the implementation of these actions;
- To meet the technical needs of the other stakeholders involved in the management of Marine and Coastal Protected Areas in the region;
- To encourage the development of the network of Mediterranean marine and coastal protected areas.
- To assist managers of each protected area so that they can manage their area as part of a network;
- To promote the research of the ecological aspects of the Adriatic sea as a whole as well as of the physical and cultural linkages between the "nodes" in the network, with a special regard to:

- collection of information on biological diversity and distribution of marine and coastal habitats;
- collection of information on ecological, socio-economic and cultural aspects of the Adriatic coasts;
- use of GIS systems in support to the decision-making processes, the management and monitoring, including the use of indicators;
- develop the concept of information hubs to disseminate and make available to the public;
- To promote and safeguard local traditions related to the management of the sea and coastal environments;
- To fund raise to achieve the goals set here.

Priority actions

The priority actions of network AdriaPAN are:
- To ensure information flow and coordination among the "nodes" of the network;
- To create and maintain an archive of all Marine and Coastal Protected Areas in the Adriatic in accordance with the methodology developed by WWF, IUCN-Med and UNEP Regional Activity Centre for Specially Protected Areas (RAC/SPA);
- To run a preliminary survey to understand whether each protected area adhering to the Charter was established through participatory process and verify whether the management is done according to commonly shared protocols and initiatives aimed an engaging local stakeholders;
- To map and evaluate the governance and management of each marine Protected area in the Adriatic;
- To develop projects and fund raise.

Those undersigning this Cerrano Charter are committed to establishing a structure that will coordinate, guide and promote the integrated system of Marine and Coastal Protected Areas in the Adriatic in the form of an association.

The number of institutions that can be part of AdriaPAN network can be expanded without geographical restrictions, as long as the objective of promoting sustainable development through participatory process is ensured.

**Opened to the subscription 26th of September 2008
Porto Caleri di Rosolina (Rovigo-Ita) Regional Park "Delta del Po Veneto".**

Bibliografy

AA.VV., *Cerrano Ieri e Oggi*, Amministrazione Provinciale di Teramo, Teramo 1983.

AA.VV., *Dalla Valle del Piomba alla Valle del basso Pescara*, Documenti dell'Abruzzo Teramano, Fondazione Cassa di Risparmio di Teramo, CARSA Edizioni, Pescara 2001.

ABDULLA A., GOMEI M., MAISON E. e PIANTE C., *Status of Marine Protected Areas in the Mediterranean Sea*, IUCN Malaga e WWF France, Gland-SUI 2008.

AGNESI S., DI NORA T., MO G., TUNESI L., *Esperienza metodologica di analisi dei dati per lo studio della nautica da diporto nell'area marina protetta di Capo Carbonara*. Biol. Mar. Mediterr. 13(1). 2006. pp.672-676.

ALBERTAZZI B. e TREZZINI F., *Gestione e tutela delle acque dall'inquinamento*, EPC libri, Roma 1999.

ARIMONDO M., LUSETTI M., MINARDI E., *Natura e Loisir*, Franco Angeli, Milano 1998.

AUGE' M., *Il bello della bicicletta*, Bollati Boringhieri, Torino 2009.

BALLINGER P., *La frantumazione dello spazio adriatico*, in COCCO E. e MINARDI E. (a cura di), *Immaginare l'adriatico. Contributi alla riscoperta sociale di uno spazio di frontiera*, Franco Angeli, Milano 2009.

BARUCCI P., *Movimento turistico ed Istituzioni: una doppia crisi*, in ENIT-ISTAT, *Rapporto sul Turismo Italiano 2004-2005*, Firenze 2004.

BECHERI E., *I turismi*, in Mercury-ENIT-ISTAT, *Rapporto sul Turismo Italiano 2004-2005*, Firenze 2004.

BENEVOLO L., *Storia dell'Architettura Moderna*, Laterza, Bari 1990.

BENCARDINO F. e PREZIOSO M., *Geografia del Turismo*, McGraw-Hill, Milano 2007.

BERTAMI F., *Coste da salvare*, in: «Costruire», n.170 luglio-agosto 1997, Roma 1997.

BIZZARRI C., *Strumenti economici per l'internalizzazione dei costi ambientali*, in CARDINALE B. (a cura di), *Mobilità traffico urbano e qualità della vita politiche e dinamiche territoriali*, Franco Angeli, DEST 304.3, Milano 2004.

BORACCHIA V. e PAOLILLO P.L. (a cura di), *Territorio sistema complesso*, Franco Angeli, Milano 1993.

BRAUDEL F., *The Mediterranean and the Mediterranean world*, Vol.I, Harper Perennial, New York, 1972.

BRUGNOLO I. e SANNA G., *Simulimpresa. L'esperienza trentina nella formazione professionale*, Franco Angeli, Roma 2008.

BRUNDTLAND G.H., *Our Common Future, Report World Commission on Environment and Development*, UNEP- Oxford University Press, London-UK 1987.

CAFFIO F., *Glossario di Diritto del Mare*, Rivista Marittima 2006, in: www.marina.difesa.it/editoria (20.12.2008).

CALZOLAIO A., *Il Piano del Parco*, Ricerche & Redazioni, Teramo 2007.

CANU A., *Rapporto sulle Aree Protette*, WWF, Roma 2006.

CARDINALE B. (a cura di), *Sviluppo Glo-cale e società nei Paesi del sistema Adriatico*, Atti Convegno Internazionale 9-11 giugno 2004 Teramo, Società Geografica Italiana-Università di Teramo-Università "G.D'Annunzio" Chieti-Pescara, Memorie S.G.I. Volume LXXVII, Roma 2006.

CARDINALE B., *Tourism and Regionalisation. Environmental, Tourist and Cultural Districts in the "Parco Gran Sasso-Monti della Laga"*, in «ANALELE UNIVERSITĂŢII DIN ORADEA», Serie GEOGRAFIE, Editura Universităţii Din Oradea, Tom XVII, 2007.

CARDINALE B., *I Distretti Ambientali Turistico Culturali del Parco nazionale Gran Sasso e Monti della Laga*, in: MAURO G., *Economia della Provincia di Teramo*, Franco Angeli, Milano, 2008.

CARDINALE B., *Mobilità delle merci e sostenibilità urbana, dinamiche territoriali e politiche di intervento*, Pàtron Editore, Bologna, 2009.

CARGINI D., MOSCA F., NARCISI V., CALZETTA A., TISCAR P.G., *Valutazione dello stato dello stock di Chamelea gallina (L. 1758) nel tratto di mare antistante l'istituenda Area Marina Protetta "Torre di Cerrano" (Teramo, Italia)*, Quaderno n°2, Atti del Progetto OASIS, NPPA Interreg-Cards/Phare "O.A.S.I.S."cod.112, Teramo 2008.

CATTANEO VIETTI R. e TUNESI L., *Le Aree Marine Protette in Italia, problemi e prospettive*, Aracne, Roma 2007.

CERUTI G., *Aree Naturali Protette. Commentario alla legge n.394/1991. Documenti*, Editoriale Domus, Milano 1993.

CESTARI M., *Genius Loci, la radice dei turismo sostenibile*, Maschietto Editore, Firenze 2007.

CESTARI M., *Genius Loci, dall'identità locale al marketing turistico*, in VALLAROLA F. (a cura di), *Aree Protette Costiere e Marine*, EditPress, AIDAP- AMP Torre del Cerrano- AMP Miramare, Teramo 2009.

CIVITARESE MATTEUCCI S., *Governo del territorio e ambiente*, in ROSSI G. (a cura di), *Diritto dell'Ambiente*, Giappichelli Editore, Torino 2008.

CNR-ISMAR, *Miglioramento e tutela della qualità dell'ambiente marino*, Quaderni di OASIS n.12, NPPA Interreg- Cards/Phare "O.A.S.I.S."cod.112, Teramo 2008.

COCCO E., *Metamofosi dell'Adriatico Orientale*, Homeless Book, Faenza, 2001.

COCCO E. e MINARDI E. (a cura di), *Immaginare l'adriatico. Contributi alla riscoperta sociale di uno spazio di frontiera*, Franco Angeli, Milano 2009.

COLETTI R., *La cooperazione transfrontaliera in Europa come strumento di governance multilivello delle aree di frontiera*, in SCARPELLI L. (a cura di), *Organizzazione del Territorio e governance multilivello*, Patron Editore, Bologna 2009.

CONTI S., DE MATTEIS G., LANZA C., NANO F., *Geografia dell'Econmia Mondiale*, UTET, Novara 2006.

CORI B. e LEMMI E. (a cura di), *Spatial Dynamics of Mediterranean Coastal Regions*, Paron Editore, Bologna 2002.

CORTE dei CONTI -Sezione di controllo per la regione siciliana-, *Relazione sull'esito dell'indagine sulla gestione delle Aree Marine Protette*, Testo presentato alla Sezione di Controllo di Palermo il 18 ottobre 2007, Conv.n.68/2007/Contr. del 2 ottobre 2007.

CRESA, *Il Turismo in Abruzzo*, Regione Abruzzo, L'Aquila, 1995.

CRESA, *Il Distretto della Strada Maestra*, PNGSL-CRESA, L'Aquila 2007.

D'AMORE F., PETRILLO P.L. e SEVERINO F. (a cura di), *Ambiente, turismo e competitività sostenibile. Come rendere la tutela ambientale un moltiplicatore di sviluppo locale*, Ricerca I-com – Ministero Ambiente e Tutela del Territorio e del Mare, Rubettino Editore, Roma 2009.

DE ASCENTIIS A., *Le regine delle Dune: Guida alla piante vascolari del litorale di Pineto*, Provincia di Teramo-WWF Italia, Teramo 2005.

DE BERNARDI A. e GANAPINI L., *Storia d'Italia 1860–1995*, Mondadori, Milano 1996.

DE MARCHI B., PELLIZZONI L., UNGARO D., *Il rischio ambientale*, Il Mulino, Bologna2001.

DI LORETO U., *Impenditoria no profit nelle aree marine protette*, in VALLAROLA F. (a cura di), *Aree Protette Costiere e Marine*, EditPress, AIDAP- AMP Torre del Cerrano- AMP Miramare, Teramo 2009.

DI NORA T., AGNESI S., TUNESI L., *Planning of marine protected areas: useful elements to identify the most relevant scuba-diving sites*. Rapp. Comm. int. Mer Médit., 38: 665, 2007.

DI PLINIO G., *Diritto pubblico dell'ambiente e aree naturali protette*, Utet libreria, Torino 1994.

DI PLINIO G., *Tavola Rotonda*, in: CARDINALE B. (a cura di), *Sviluppo Glo-cale e società nei Paesi del sistema Adriatico*, Atti Convegno Internazionale 9-11 giugno 2004 Teramo, Società Geografica Italiana-Università di Teramo-Università "G.D'Annunzio" Chieti-Pescara, Memorie S.G.I. Volume LXXVII, Roma 2006.

DI NORA T. e AGNESI S., *Supporto decisionale per le aree marine protette mediante GIS*, in VALLAROLA F. (a cura di), *Aree Protette Costiere e Marine*, EditPress, AIDAP- AMP Torre del Cerrano- AMP Miramare, Teramo 2009.

DIVIACCO G., *Aree Protette Marine, finalità e gestione*, ComunicAzione Edizioni, Forlì 1999.

FERRAJOLO O., *Le aree specialmente protette del Mediterraneo*, in MARCHISIO S. (a cura di), *Codice delle Aree Protette*, Giuffrè Editore, Milano 1999

FERRAJOLO O., *Le aree protette marine tra obblighi internazionali diritto italiano*, in GRAZIANI C.A. (a cura di), *Un'utopia istituzionale: le aree naturali protette a dieci anni dalla legge quadro*, Giuffrè Editore, Milano 2002.

FONDERICO F., *Informazioni- Parte III- Introduzione*, in ROSSI G. (a cura di), *Diritto dell'Ambiente*, Giappichelli Editore, Torino 2008.

FRANCHINI D. (a cura di), *Pianificazione delle Aree Costiere*, CISIAC- Guerini&Associati, Milano1988.

FRANZOSINI C., *L'efficacia di gestione delle Amp come sistema di verifica e coordinamento internazionale*, in VALLAROLA F. (a cura di), *Aree Protette Costiere e Marine*, EditPress, AIDAP- AMP Torre del Cerrano- AMP Miramare, Teramo 2009.

GAMBINO R., *Parchi Naturali*, NIS Nuova Italia Scientifica, Roma 1991.

GEMMITI R. e CONTI PUORGER A., *Governo, governance, sussidiarietà e territorio*, in SCARPELLI L. (a cura di), *Organizzazione del Territorio e governance multilivello*, Patron Editore, Bologna 2009.

GON D., *Verso Est uno sguardo alla geopolitia dell'Adriatico*, in COCCO E. e MINARDI E. (a cura di), *Immaginare l'adriatico. Contributi alla riscoperta sociale di uno spazio di frontiera*, Franco Angeli, Milano 2009.

GRANZOTTO A., LIBRALATO S., RAICEVICH S., GIOVANARDI O., PRANOVI F. (a cura di), *Analisi dello stato delle risorse alieutiche dell'alto Adriatico mediante le serie storiche di sbarcato*, in: Biologia Marina Mediterranea, n.13 (1). 2006.

GRECO N., *Le Aree marine protette nel quadro della Gestione integrata delle coste*, in ROSSI Gianluca (a cura di), *Diritto dell'Ambiente*, Giappichelli Editore, Torino 2008.

GIACOMINI V. e ROMANI V., *Uomini e Parchi*, Franco Angeli, Milano 1982.

GUIDETTI P., MILAZZO M., BUSSOTTI S., MOLINARI A., MURENU M., PAIS A., SPANO' N., BALZANO R., AGARDY T., BOERO F., CARRADA G., CATTANEO-VIETTI R., CAU A., CHEMELLO R., GRECO S., MANGANARO A., NOTARBARTOLO DI SCIARA G., FULVIO RUSSO G., TUNESI L., *Italian marine reserve effectiveness:Does enforcement matter?*, in Biological Conservation n.141, Elsevier Ltd., 2008. pp.699-709.

HOCKINGS M., STOLTON S., LEVERINGTON F., DUDLEY N., COURRAU J., *Evaluating Effectiveness: A framework for assessing management effectiveness of protected areas*, 2nd edition. IUCN, Gland, Switzerland and Cambridge-UK 2006.

IANNI V. e TOIGO M., *L'impegno della Regione Marche per la solidarietà e la cooperazione internazionale*, Regione Marche- Movimondo, Ancona 2002.

ICRAM, *Qualità degli Ambienti Marini Costieri Italiani 1996-1999*, Ministero Ambiente-Servizio Difesa Mare, Roma 2000.

IELARDI G., *Proteggiamo la natura eredità del futuro*, Federparchi, Roma 2008.

ILLCH I., *Energie et Equitè*, in LA CECLA F. (a cura di), *Elogio della bicicletta*, Bollati Boringhieri, Torino 1973.

IUCN-World Commission on Protected Areas, *Establishing Marine Protected Area Networks- Making It Happen*, IUCN-WCPA, National Oceanic and Atmospheric Administration, The Nature Conservancy, Washington D.C.-USA 2008.

JOVANOVIC B., *Tra urbano e rurale: lo sviluppo del turismo sostenibile nella regione adriatico meridionale e ionica*, in COCCO E. e MINARDI E. (a cura di), *Immaginare l'adriatico. Contributi alla riscoperta sociale di uno spazio di frontiera*, Franco Angeli, Milano 2009.

LABATUT B., *Acquacoltura*, FAO 2008, in: www.fao.org/news (06.10.2008).

MAMMARELLA L. (1993), *Piazzeforti e Torri Costiere*, Borgia Editore, Roma.

MARCHISIO S., DELLA FINA V., FERRAJOLO O., SALBERINI G., TAMBURELLI G., *Codice delle Aree Protette*, Giuffrè Editore, Milano 1999.

MASSIMI G., *Movimenti virtuali di popolazione (1861-1991). La componente altimetrica*, in: LANDINI P. (a cura di), *Abruzzo. Un modello di Sviluppo Regionale*, Società Geografica Italiana, Roma 1991.

McINTYRE G., *Sustainable Development, Guide for Local Planners*, World Tourist Organization, Madrid-ESP 1993.

MATTM, *Il manuale del buon diportista in barca nelle aree protette*, Ministero Ambiente Territorio e Tutela del Mare-UCINA-Legambiente, Roma 2008.

MATTM, *Valutazione dell'efficacia di gestione delle Aree Marine Protette italiane*, Ministero dell'Ambiente e Tutela del territorio e del Mare- WWFItalia-Federparchi, EUT Edizioni Università Trieste, Trieste 2007.

Mercury-ENIT-ISTAT, *Rapporto sul Turismo Italiano 2004-2005*, Firenze 2004.

Mercury- Osservatorio Parlamentare Turismo- Ministro del Turismo, *Rapporto sul Turismo Italiano 2008-2009*, Franco Angeli, Milano 2009.

MINARDI E., *Verso l'Euroregione adriatica. Le azioi possibili per la cooperazione e lo sviluppo della regione*, in COCCO E. e MINARDI E. (a cura di), *Immaginare l'adriatico. Contributi alla riscoperta sociale di uno spazio di frontiera*, Franco Angeli, Milano 2009.

MOSCHINI R. (a cura di), *Aree protette e nautica sostenibile*, Edizioni ETS, Pisa 2009.

MOSCHINI R., *Le istituzioni e la gestione delle aree protette*, Numero speciale di quaderni del Parco di Migliarino San Rossore Massaciuccoli, Centro Studi Valerio Giacomini, Suppl. Toscana Parchi n.8, Pisa 2004.

MOSCHINI R., *Parchi, a che punto siamo?*, Edizioni ETS, Pisa 2006.

NAVIGLIO L., *Strumenti per la gestione sostenibile di aree costiere*, in VALLAROLA F. (a cura di), *Aree Protette Costiere e Marine*, EditPress, AIDAP- AMP Torre del Cerrano- AMP Miramare, Teramo 2009.

NASTI A. e MARINO D., *Aree marine protette e pesca: alla ricerca della governance*, in Atti del Convegno Siracusa 5 dicembre 2008, *Aree marine protette e Pesca: alla ricerca delle buone pratiche condivise*, ANFE Sicilia- ARPA Sicilia- LANDS Onlus, Siracusa 2009.

NOTARBARTOLO di SCIARA G., *Foreword*, in: PIANTE C., ABDULLA A., GOMEI M., MAISON E., *Status of Marine Protected Areas in the Mediterranean Sea*, IUCN Malaga e WWF France, Gland-SUI 2008.

OLIVIERI V., *L'attività nautica e di pesca e l'ecosistema marino: fattori di minaccia alla conservazione di Cetacei e Tartarughe*, in VALLAROLA F. (a cura di), *Aree Protette Costiere e Marine*, EditPress, AIDAP- AMP Torre del Cerrano- AMP Miramare, Teramo 2009.

OCSE, *Atto Unico Europeo*, Parigi-FRA 1975.

ORT-Osservatorio Regionale sul Turismo, *Dati di sintesi sul turismo regionale 1999*, Regione Abruzzo, Pescara 2000.

PARLAMENTO ITALIANO, COMMISSIONE VIII-Ambiente, territorio, lavori pubblici, *Sistema di gestione amministrativa degli Enti Parco nazionali*, Atti parlamentari XIV legislatura, Indagini conoscitive e documentazioni legislative n.13, Camera dei Deputati, Roma 2004.

PERSI P., *Territori Contesi, campi del sapere, identità locali, istituzioni, progettualità paesistica*, Istituto Interfacoltà di Geografia Università di Urbino "Carlo Bo"- AIIG Marche - Comune di Pollenza (Mc), Urbino 2009.

PIANTE C., ABDULLA A., GOMEI M., MAISON E., *Status of Marine Protected Areas in the Mediterranean Sea*, IUCN Malaga e WWF France, Gland-SUI 2008.

POLCI S., GAMBASSI R., *Un turismo di prossimità (anche culturale?)*, in ENIT-ISTAT, *Rapporto sul Turismo Italiano 2004-2005*, Firenze 2006.

RAICEVICH S., FORTIBUONI T., GIOVANARDI O., *Integrazione di fonti storiche, statistiche, antropologiche e scientifiche per individuare l'estirpazione di specie marine minacciate nel Mediterraneo*, in GERTWAGEN R., RAICEVICH S., FORTIBUONI T. e GIOVANARDI O. (a cura di), *Il mare, Com'era, Le interazioni tra uomo e ambiente nel Mediterraneo dall'Epoca Romana al XIX secolo*, Atti II Workshop Internazionale HMAP del Mediterraneo e Mar Nero, Chioggia 27-29 settembre 2006, Supplemento ai quaderni ex-ICRAM, ISPRA Chioggia, Venezia 2008.

RAVAZZA N., *Il sale e il sangue, storie di uomini e tonni*, Magenes, Milano 2007.

RELINI G. (a cura di), *Dominio Pelagico, i Santuario dei cetacei Pelagos*, MATTM-Museo Friulano Storia Naturale, Quaderni Habitat, Udine 2007.

RHI SAUSI J.L., COLETTI R., CUGUSI B., *Strumenti e metodologie dei programmi di prossimità nel Mediterraneo nella fase di transizione. Prospettive per la cooperazione interregionale*, Research Paper, in atti della Conferenza Napoli 22-23 luglio 2004, *Sperimentazione delle politiche di prossimità nel Mediterraneo occidentale*, www.cespi.it (luglio 2008).

RODI S., *In 15 anni il cemento ha ricoperto un'area grande come Lazio e Abruzzo* in «Il Corriere della Sera», 4 agosto 2009, Torino 2009.

ROMANI V., *Il Paesaggio, Teoria e Pianificazione*, Franco Angeli, Milano 1994.

ROMANO B., *Oltre i Parchi, la rete verde regionale*, Andromeda Editrice, L'Aquila 1996.

ROLLI G.L. e ROMANO B., *Progetto Parco*, Università degli Studi dell'Aquila, Andromeda Editrice, Colledara (Te) 1995.

ROTTA A., *I partenariati territoriali nello spazio adriatico. Origine, evoluzione, prospettive*, in COCCO E. e MINARDI E. (a cura di), *Immaginare l'adriatico. Contributi alla riscoperta sociale di uno spazio di frontiera*, Franco Angeli, Milano 2009.

ROVITO C., *Le Aree Marine Protette: divieti, approfondimenti e decisioni della Corte di Cassazione*, Diritto dell'Ambiente, testata giornalistica on line: www.dirittoambiente.com (dicembre 2009).

SCARPELLI L. (a cura di), *Organizzazione del Territorio e governance multilivello*, Patron Editore, Bologna 2009.

SCOTT J., TRUBEK D.M., *Mind the Gap: law and new approach to governance in the European Union*, European Law Journal, Volume 8, n.1, March 2002.

SEGRE A., DANSERO E., *Politiche per l'ambiente. Dalla natura al territorio*, UTET, Torino 1996.

SIMEON M.I., *Il turismo dei beni culturali*, in ENIT-ISTAT, *Rapporto sul Turismo Italiano 2004-2005*, Firenze 2006.

SIMONCINI A., *Ambiente e protezione della natura*, Università di Firenze-Facoltà di Economia, CEDAM Padova 1996.

SPALDING M.D., FOX H.E., ALLEN G.R., DAVIDSON N., FERDAÑA Z.A., FINLAYSON M., HALPERN B.S., JORGE, M.A., LOMBANA A., LOURIE S.A., MARTIN K.D., MCMANUS E., MOLNAR J., RECCHIA C.A., ROBERTSON J., *MarineEcoregions of the World: A Bioregionalization of Coastal and Shelf Areas*, BioScience 57 (7-8). 2007 pp.573-583.

SPOTO M., *AdriaPAN: un nuovo progetto per le aree protette marine e costiere dell'Adriatico*, in VALLAROLA F.(a cura di), *Aree Protette Costiere e Marine*, EditPress, AIDAP- AMP Torre del Cerrano- AMP Miramare, Teramo 2009.

STOCCHIERO A., *I partenariati territorili nella politica di prossimità*, Research Paper CeSPI. www.cespi.it (luglio 2004).

TASSI F., *Esperienze di gestione di parchi nazionali*, intervento alla Tavola Rotonda del WWF Italia tenutasi il 23 novembre 1992 all'Abbadia di Fiastra (Mc) in atti WWF Italia, *Metodologia di analisi ed ipotesi di zonizzazione per un parco Nazionale*, Biemmegraf, Macerata 1994.

TISCAR P.G. (a cura di), *Indagine conoscitiva sull'Area Marina Protetta "Torre del Cerrano"*, Università degli Studi di Teramo -Facoltà di Medicina Veterinaria, Teramo 2001.

TREU M.C., *Riferimenti e ipotesi per una procedura di pianificazione ambientale*, in BORACCHIA V., PAOLILLO P.L. (a cura di), *Territorio sistema complesso*, Franco Angeli, Milano 1993.

TALLONE G., *I Parchi come Sistema*, Edizioni ETS, Pisa 2007.

TRANQUILLI LEALI R. e LO BOSCO L., *La compatibilità dell'esercizio dell'attività di pesca con la tutela dell'ambiente marino anche in relazione all'istituzione delle Aree Marine Protette*, in Quaderni OASIS n.1, Provincia Teramo- Università Teramo, NPPAInterreg-Cards/Phare "O.A.S.I.S.", Teramo 2008.

TUNESI L., DI NORA T., AGNESI S. (2004), *Potenzialità delle Aree marine protette per la gestione delle risorse ittiche nelle acque italiane*. Biologia Marina Mediterranea, 11 (2): 33-39.

TUNESI L., AGNESI S., DI NORA T., *La gestione del turismo subacqueo nelle aree marine protette (AMP): gli elementi prioritari*. Atti del workshop internazionale Ostia, Roma, 17 e 18 Febbraio 2005 *"Le attività subacquee nelle aree marine protette e gli impatti sull'ambiente: esperienze mediterranee a confronto"*, Palombi Editori, Roma 2007.

TUNESI L., *La ricerca a supporto delle Aree Marine Protette in Italia*, in VALLAROLA Fabio (a cura di), *Aree Protette Costiere e Marine*, EditPress, AIDAP- AMP Torre del Cerrano- AMP Miramare, Teramo 2009.

TUNESI L., *Biodiversità e Are marine protette, un network per un sistema nazionale, europeo e mediterraneo*, intervento Convegno internazionale Bari il 29 gennaio 2010, *"2010 Anno Internazionale della Biodiversità: ora è il tempo di agire"*, Mediterre, Bari 27-30 gennaio 2010. (www.greenreport.it 01.02.2010).

UNEP-WCMC, *National and Regional Networks of Marine Protected Areas: a review of progress*, UNEP World Conservation Monitoring Centre, Cambridge-UK 2008.

VALLAROLA F., *Gran Sasso Monti della Laga, il Parco Nazionale*, MEDIA Edizioni, Teramo 1998.

VALLAROLA F., *Cerrano, terre da proteggere*, Ricerche&Redazioni, Teramo 2005.

VIALE G., *Vita e morte dell'automobile. La mobilità che viene*, Bollati Boringhieri, Torino 2007.

VIALE G., *Muoversi sostenibilmente* in «Piemonte Parchi», n.187, Torino 2009.

VALLAROLA F. (a cura di), *Aree Protette Costiere e Marine*, EditPress, AIDAP-AMP Torre del Cerrano- AMP Miramare, Teramo 2009.

VALLAROLA F. (a cura di), *Le Aree Marine Protette*, Edizioni ETS, Pisa 2011.

WWF, *Ecoregion action programmes: a guide for practitioners*, Roma 2003.

ZOPPI C., *Aree protette marine e costiere, questioni di pianificazione del territorio*, Gangemi Editore, Roma 1993.

www.ingramcontent.com/pod-product-compliance
Lightning Source LLC
Chambersburg PA
CBHW072212170526
45158CB00002BA/566